Lab Manual to accompany

Home Technology Integration Fundamentals and Certification

Cisco Learning Institute

and

Academic Business Consultants

PEARSON

Prentice
Hall

Upper Saddle River, New Jersey

Columbus, Ohio

Editor in Chief: Stephen Helba
Assistant Vice President and Publisher: Charles E. Stewart, Jr.
Production Editor: Kevin Happell
Design Coordinator: Diane Ernsberger
Cover Designer: Tim Hunter
Production Manager: Matthew Ottenweller
Marketing Manager: Ben Leonard

This book was set in Times Roman by Prentice Hall. It was printed and bound by Courier/Kendallville.
The cover was printed by Phoenix Color Corp.

Pearson Education Ltd.
Pearson Education Singapore Pte. Ltd.
Pearson Education Canada, Ltd.
Pearson Education—Japan

Pearson Education Australia Pty. Limited
Pearson Education North Asia Ltd.
Pearson Educación de Mexico, S.A. de C.V.
Pearson Education Malaysia Pte. Ltd.

10 9 8 7 6 5 4 3 2 1
ISBN 0-13-142108-5

Preface

Dozens of individuals participate in the construction or retrofitting of a home or small commercial building. Electricians, cabling technicians, security installers, low-voltage experts, and home entertainment installers all participate in the development of the modern home's "smart" installations. The convergence of these fields demands the skills of a new type of technician—one that can "integrate" all of these diverse technologies into one systematic deliverable solution. One of the best ways to build technical skills is to participate in industry-based, hands-on labs. This lab manual is designed to help students manage the lab activities.

The Cisco Learning Institute is a 501(C)(3) public benefit corporation founded and originally funded by Cisco Systems. The Institute is dedicated to enhancing the way that students learn and teachers teach using technology. Using the basics of a curriculum provided by Cisco Systems, the Cisco Learning Institute developed a complete online curriculum designed to direct students' learning. Academic Business Consultants, Inc. helped with the authoring of the original material. Leviton Manufacturing provides the lab configuration that is so critical to the hands-on aspect of the program. HAI and Premise provide the industry-leading software that helps to control the devices in the lab and home. BlueVolt provides students around the world with online hosting services for the curriculum and the assessment system. All of these organizations have worked very hard to provide the HTI industry and its students with the very best in industry-based training. For more information, visit these websites:

http://hti.ciscolearning.org	http://www.bluevolt.com
http://www.comptia.org	http://www.prenhall.com
http://www.internethomealliance.com	http://www.homeauto.com
http://www.leviton.com	http://www.premisesystems.com

INTENDED AUDIENCE

This book and its companion curriculum and labs are intended to serve the needs of individuals interested in understanding the field of home technology integration. The material presented in these training products is designed to provide students with the background knowledge, hands-on experience, and overall confidence to prepare for the CompTIA HTI+ exams and (more importantly) for a solid career in a growing and exciting field.

All of the HTI+ exam objectives are covered in this text, in the companion lab manual, or in the online curriculum. The authors assume that the users of these products have some knowledge of electricity.

SAFETY FIRST

The activities covered in this program deal with both high- and low-voltage electricity applications. The authors and the publishers of this material urge all students and teachers to exercise great caution when dealing with electricity. Although the labs in this curriculum have been carefully tested, and all of the training sites have been given careful directions for constructing the lab walls, each individual should exercise great caution with the lab equipment. Electrical shock can cause injury or death.

THE CISCO LEARNING INSTITUTE HOME TECHNOLOGY INTEGRATION TRAINING PROGRAM

The Cisco Learning Institute Home Technology Integration Training Program is a carefully designed total training solution that includes a textbook, a lab manual, hands-on labs, and an online curriculum. All of these components are engineered to work together to provide the best possible teaching and learning experience.

Textbook

The textbook provides the student and the teacher with a completely portable reference to the entire curriculum. The text includes detailed objectives for each chapter, as well as chapter summaries, key terms, and review questions. The text is a complete reference to the curriculum and provides the basis for independent study and HTI+ exam preparation. Also packaged with the main text is an examGEAR CD containing hundreds of questions that will help readers prepare for the exam.

Lab Manual

The lab manual, which is part of the print bundle, includes documentation for all fifty-one labs. The manual provides students with a ready reference to use with the labs in class and a convenient place to record lab activities. The labs are designed to give students a true view of the situations that they will encounter in the field. The equipment used in the labs is manufactured by Leviton Manufacturing and HAI, two of the largest and best-known manufacturers of low-voltage home automation devices. This manual is designed to offer an easy-to-follow guide for completing the labs and recording results.

Certification Kit

The Certification Kit on the accompanying CD provides a variety of tools that should prove useful to students preparing for the HTI+ exam. The CD includes a variety of student activities and simulations that provide work-based practice for the technician. In addition, supplemental test questions have been added to provide a rich practice environment for the HTI+ exam.

Hands-on Labs

The goal of this course is to help produce knowledgeable technicians who are capable of installing and servicing complex integrated systems within the home. No amount of reading can replace the value of hands-on experience. The labs use state-of-the-art tools and equipment manufactured by industry leaders. Leviton equipment is used for the labs dealing with low-voltage structure wiring, and audio and video installations (and for most of the security installations). HAI and Premise software are used for the control and programming labs.

Online Curriculum

The full curriculum is available online from either a local training site or from BlueVolt, our national provider. The curriculum includes text, illustrations, and many exciting simulations and interactivities designed to foster understanding and build confidence. The curriculum is available twenty-four hours a day and can be accessed from any computer connected to the Internet.

ACKNOWLEDGMENTS

The Cisco Learning Institute wants to thank several organizations and individuals for their help and support during the developmental process: Bridget Bisnette, Cisco Systems; Ian Hendler, Leviton Manufacturing; Bruce Luie, Premise Systems; Lisa Bourdeaux, BlueVolt; Jay McLellan, HAI; and David McLean, Fluke Networks.

Many individuals worked on the creation of the program. Academic Business Consultants did an excellent job of managing the authoring and development process. Chuck Stewart provided craft and skill to the manuscript as developmental editor. Nancy Sixsmith did a superb job as our copyeditor. Our editor at Prentice Hall, Charles Stewart, did an excellent job guiding the project. Thanks also to our production editor, Alex Wolf, who provided outstanding production coordination.

We are also indebted to the following technical reviewers: Ron Kovac, Ball State University; Russell Hillpot, Gaddsdon State Community College; Dennis Quatrine, Henry Ford Community College; and Tom Moffses, North Florida Community College.

Thanks to all of our students and instructors. We hope that this program will provide you with the skills to enter this exciting field.

Dave Dusthimer
Publisher, Cisco Learning Institute

Table of Contents

1-1	Installation Tools	1
3-1	Category 5 Cabling	3
3-2	Installing a Modem	13
3-3	Installing a Network Interface Card	17
3-4	Networking PCs	27
3-5	Configuring the Leviton Residential Gateway	31
3-6	Setting Up a Firewall	37
3-7	Configuring Ports	41
3-8	Home Networking Using NetBEUI	45
4-1	Telecom Demarc	51
4-2	Line Distribution	57
4-3	Using an RJ31X	61
5-1	Exploring Coaxial Cable	63
5-2	RG-6 Connectors	67
5-3	Video Adapters	71
5-4	RF Distribution	75
5-5	Video Splitters and Combiners	81
5-6	Video Camera Installation	83
5-7	RF Amplifiers	89
5-8	DC Power Distribution	91
5-9	Decora Media System	95
5-10	Audio Distribution	99
5-11	Speaker Loads	105
5-12	Infrared Systems	111
6-1	HAI Control Box Setup	117
6-2	Video Sequencer Configuration	123
7-1	HAI Thermostat Connections	127
7-2	HAI System Power-Up	133
7-3	Programming X-10 Lighting Controls	139
7-4	Addressing Dimmers	141
7-5	HAI Remote Thermostat Control	143
7-6	Motion Detectors	149
8-1	Connectors	153
9-1	Measuring AC	157
9-2	Measuring DC	159
10-1	HAI X-10 Controls	165
10-2	HAI Software Installation	169
11-1	Understanding Blueprints	181
12-1	Configuring the Structured Media Center	187
12-2	Door Facts	197
13-1	Installing Jacks	199
13-2	Work Area Outlets	203
14-1	Testing Category 5e Cables	205
14-2	Construction-Related Faults	209
15-1	CATV for Multiple Dwelling Units	211
15-2	Wireless Communications	213

Installation Tools

INTRODUCTION

The purpose of this lab is to identify, examine, and demonstrate how to safely use common tools during cabling installations. It is important to determine which tools will be needed to install, test, and troubleshoot cable installation jobs. In order to ensure quality installations and facilitate optimal safety, it is critical that the installer be aware of the tools that will be required and of how to prevent injury in the workplace. This lab will assist with mastering the techniques for proper cable termination, testing, and identifying how the proper use of tools will prevent accidents.

This lab should be done when the instructor is present, and care should be taken to identify any safety measures associated with the tool. It is recommended that safety glasses be worn when using the equipment in the classroom and in the field.

WHAT YOU WILL DO

▸   Identify common tools used in the cabling field.

▸   Examine the features, benefits, and safety measures of each tool.

▸   Demonstrate how to handle the tools appropriately.

WHAT YOU WILL NEED

▸   Tone probe

▸   All-in-one telephone tool

▸   KT 8 cutter

▸   SealTite 59/6 waterproof CATV F crimp tool

▸   Coax stripper

▸   LAN Pro Navigator

▸   Data SureStrip

▸   Electrician's scissors

▸   SurePunch Pro impact termination tool

Figure 1-1-1 illustrates a tool belt with common HTI tools.

Figure 1-1-1 Tool belt and common HTI tools

Inspect and touch all tools to identify the features, benefits, and safety practices for each.

Tone Probe

The cable tracer (amplified probe) shown in Figure 1-1-2 features a volume control and a signaling LED that can be used to verify the tone on the pair. The tone probe cable tracer tests the continuity and signal of installed wire and cable through walls and through thick cable insulation. It will quickly and easily identify a cable pair carrying a tone signal, and identify patch and termination points of the wire. The probe functions in three different operating modes: speaker and light, speaker only, or light only. Two advantages to using this tool are that light mode uses gain control circuitry and operates when it receives true signal, not just random noise; and this tool allows the installer to work in noisy office environments during normal hours without disturbing employees.

Figure 1-1-2 Amplified probe

All-In-One Telephone Tool

This crimping tool cuts, strips, and crimps modular plugs (see Figure 1-1-3). It is easy to use and is a convenient tool that makes most telephone and datacom jobs very fast. This tool cuts and strips flat telephone cable and crimps 6P4C/6P6C (RJ-11/RJ-12) modular telephone plugs and 8P8C (RJ-45) data plugs, which ensure a good crimp every time. When the ratchet cycle is complete, the crimp is done; there no need to press any harder than when completing the ratchet cycle. Care should be taken when using this tool because the razor-sharp blade can cause injuries. Be careful not to place your fingers inside the tool during a crimp.

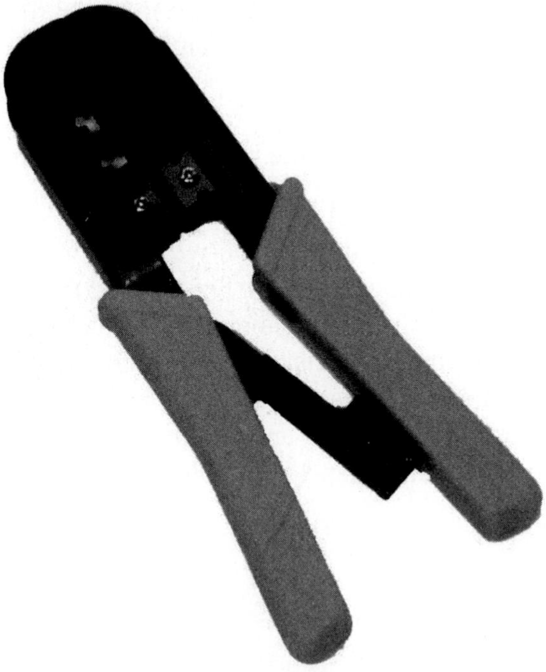

To strip flat satin cable:

1. Open the handles of the tool completely.

2. For six-conductor cable, insert the cable into the blade area. Rest the front of the cable on the small ridge located inside the handle under the blades.

3. For eight-conductor cable, insert the cable under the blades until the end protrudes from the back side of the tool.

Figure 1-1-3 All-in-one telephone tool

To crimp 6P4C connectors:

1. Attach the 6P4C connector to the cable.

2. Install the connector into the 6P location of the tool. Ensure that the connector key is facing downward to fit the contour of the crimp area.

3. Close the handles through a complete ratchet cycle until the handles open. The crimp is complete.

To crimp 8P8C connectors:

1. Attach the 8P8C connector to the cable.

2. Install the connector into the P location of the tool. Ensure that the connector key is facing downward to fit the contour of the crimp area.

3. Close the handles through a complete ratchet cycle until the handles open. The crimp is complete.

KT 8 Cutter

The KT 8 cutter is a common tool used for cutting solid, stranded, and flexible cables up to AWG6 (16 mm). As shown in Figure 1-1-4, the blades are outlined for two conductor size ranges for precise and effortless cuts. The KT 8 cutter does not deform the cable, which keeps it round and easy to slip the connector onto. This tool is perfect for coax and data cable. Care should be taken when using this very sharp tool because the blade can cause injuries.

SealTite 59/6 CATV F Crimp Tool

This tool provides durable, all-weather connections for cable TV applications (see Figure 1-1-5). The tool performs compression crimping of water-resistant CATV F connectors. Its fully ratcheting one-cycle crimp function creates a seal and watertight connection. Keep fingers and clothing away from the tool to prevent injury and damage to clothing.

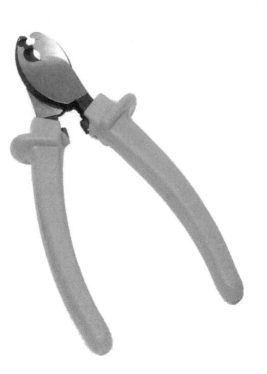

Figure 1-1-4 KT 8 cutter

Follow these steps to use this tool:

1. Strip the cable using a coaxial stripping tool. The strip should be two-level for exposure of center conductor and shield/dielectric.

2. Install the cable into the connector.

3. Open the tool handles. With the connector facing downward, clip the connector into the crimp area with the cable extending out the top of the tool. Make sure that the cable is held in place by the two upper retaining latches.

4. Close the handles through a full ratchet cycle until the plastic end snaps into the connector and seals the cable.

5. Open the handles and remove the cable assembly.

6. Push the safety release down to disengage the ratchet if needed.

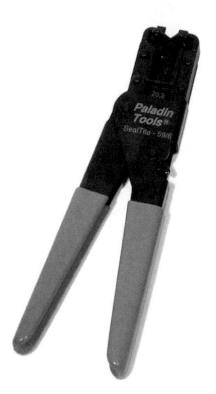

Figure 1-1-5 SealTite 59/6 waterproof CATV F crimp tool

Coax Stripper

The easy-to-use coax stripper performs two- and three- level stripping of the outer jacket, shield braid, and inner dielectric (see Figure 1-1-6). The features and benefits are that it is fully adjustable with a simple screw, and it is self-regulating to prevent nicking of the cable. Care should be taken when using this tool because the razor-sharp blade can cause injuries.

To set up blade strip lengths, follow these steps:

1. Find the blade retaining pin on the bottom jaw (blade side) and remove the pin.

2. Move the blades for the type of strip that you want. For two-level stripping, remove the center blade.

3. Close the jaws and replace the blade retaining pin.

Note: The coax stripper can also be set for two-level stripping for Series 59 and Series 6 CATV Type-F connectors.

4. Remove the center blade and replace it with the conductor blade from the far-right slot.

5. Close the jaws and replace the blade retaining pin.

To set the blade cut depth, follow these steps:

1. Using the hex key, adjust the three hex screws on the underside of the bottom jaw.

2. Adjust each hex screw to cut the required depth.

To perform the strip, follow these steps:

1. Put the cable into the jaws in the direction indicated on the top of the tool.

2. Spin the tool in the direction of the arrow located on the side of the stripper three or four times around the cable.

3. Remove the cable from the tool.

4. Pull off the stripped sections and inspect the cable for accuracy.

5. Repeat the setup of the strip length and/or depth until the desired strip is achieved.

Note: Do not hold down or force down the stripping jaws during the stripping cycle. The jaws are spring-loaded and will adjust automatically.

Figure 1-1-6 Coax stripper

LAN Pro Navigator

The LAN Pro Navigator (Data & Coaxial Network Test Set) is a basic cable tester that provides a large amount of information regarding the accuracy of cable terminations and the conditions of cables. As shown in Figure 1-1-7, it consists of two units: a control unit and a remote unit, both equipped to test UTP or STP cables, which connect to 8P8C connectors. The LAN Pro Navigator detects opens, shorts, crossed wires, and split pairs. It will also test coaxial cable for continuity and shorts. It features a one-button, simple testing (PASS/FAIL) or fault find, the capability to trace wires with tone and remote lights, in addition to generating oscillating audio tone for use with tone probes to locate cables. The main tester has an automatic power shutdown circuit to save battery life. The unit will turn off after approximately 20 seconds of nonuse during normal testing and fault check operations and after 30 minutes in tone mode. To prevent power shutdown, periodically press the Fault Check/Tone switch. Pressing the On/Test switch after a test will turn the unit off.

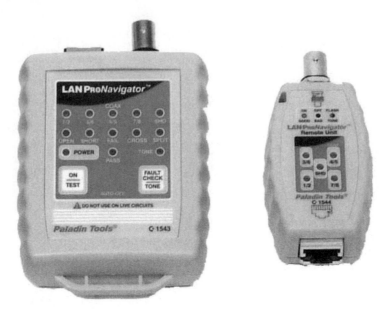

Figure 1-1-7 LAN Pro Navigator

Prior to using the LAN Pro Navigator, you should read the documentation that is included with this data and coaxial network tester.

1. To determine if you have a low battery in your tester, connect a known-good data patch cable between the main tester and remote unit.

2. Press the ON/TEST button to turn the unit on.

3. Press the ON/TEST button to initiate a test.

4. If the POWER light flashes on and off after initiating the test, the battery is low and needs to be replaced.

Warning: Never open the case of the tester. Access the battery through the battery door provided. Avoid high-moisture environments.

Data SureStrip

This tool strips and cuts round stranded or solid UTP/STP voice and data cables up to a diameter of 0.5 inches (see Figure 1-1-8). It is a universal, easy-to-use cable cutter and stripper for data, telephone, and network installations. The Data SureStrip cuts and strips all data cables, including UTP/STP data cables, 25-pair multiconductor cable, four-, six-, and eight-conductor flat satin cables, and coaxial cables (you may have to adjust for various strip depths). Its adjustable blade depth prevents damage to shield and conductor, and will also strip and cut some types of small coaxial cable and hook-up wire. Care should be taken when using this tool because the razor-sharp blade can cause injuries.

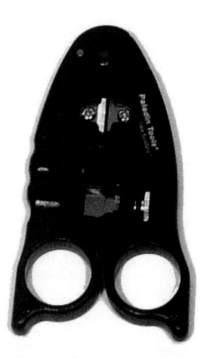

Figure 1-1-8 Data SureStrip

To cut wire and cable, follow these steps:

1. Insert the wire or cable into the cable cutter section.

2. Grip the tools and squeeze the top handle down until the article is cut.

To strip round wire and cable, follow these steps:

1. Adjust the blade depth screw to set the blade cut depth to the desired level to ensure a non-scoring strip. This may take several test settings before achieving optimum depth.

2. Place the wire/cable into the center round hole of the tool.

3. Place your finger into the finger loop and spin the tool clockwise around the cable one or two times. Do not press down on the top lever. The tool has a self-regulating spring to control stripping.

4. Open the tool and remove the cable. Pull the stripped insulation off the cable.

To strip flat satin telephone cable, follow these steps:

1. Place the cable into the flat cable stripper area located at the front of the tool.

2. Close the tool and hold it in palm of one hand, and hold the cable steady in the other hand.

3. Using a straight, no-angled motion, pull the tool away and off from the end of the cable. This will strip off the outer jacket, exposing the inner conductors.

Electrician's Scissors

These common types of scissors are also known as "snips." As shown in Figure 1-1-9, this tool is designed to cut CAT 5 cable and other types of wire during cable installation. Two notches on one side of the blade are used to skin insulation from individual conductors and can be used for scoring of cable jackets. Safety measures should be implemented when using this tool, which can pinch fingers between the handle and cause puncture wounds or cuts.

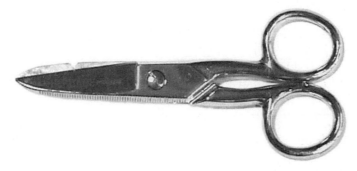

Figure 1-1-9 Electrician's scissors

SurePunch Pro Impact Termination Tool

This tool is used for wire preparation and panel maintenance (see Figure 1-1-10). It includes an adjustable impact force to meet the demands of all cables. The blades twist and lock into place and will accept other manufacturers' twist-and-lock blades. There is a latch cover blade storage area in the end of the handle that allows you to secure the blade(s). Blades are available to terminate and cut telephone and data cable into a variety of termination panels, including 66, 88, 108, 109, 110, BIX, and Krone LSA bifurcated terminals. The tool includes finger grips and a non-slip cushioned handle for comfort along with a spring mechanism that requires less hand force to actuate with adjustable HI/LO spring compression. Safety measures should be implemented when using this tool, which can cause serious injury. Its razor-sharp blades and multiuse spudger (a small hook-shaped tool used to remove conductors from terminations when they are improperly terminated), screwdriver, and IDC contact insertion tools can cause lacerations and puncture wounds.

Conclusions

During this lab, you identified the different types of tools that are necessary to complete a cabling installation job. You examined the characteristics, features, and benefits of each tool. You demonstrated the proper handling for each tool, and recognized and mastered the safety techniques associated with each.

Exercises

1. What tool would you use to terminate a UTP cable? _____

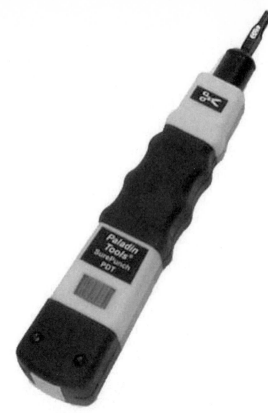

Figure 1-1-10 SurePunch Pro impact termination tool

2. What is the name of the tool that you would use to test a cable for shorts or opens? _____

3. What is the name of the tool that you would use to crimp a Type-F connector using coaxial cable? _____

4. What is the name of the tool that is used to identify patch and termination points of a wire?_____

5. Which tool would you use to terminate cable on a 110 block? _____
   _____

6. Which tools can be used to strip the conductors of a Category 5e cable? _____
   _____

7. Which tool can be used to crimp voice and data plugs (6P6C or 8P8C)? _____
   _____

Category 5 Cabling

INTRODUCTION

A Category 5 (CAT 5), Category 5e (CAT 5e), or greater unshielded twisted-pair (UTP) Ethernet network patch cable (or patch cord) will have four-pair (eight wires). In this lab, it will be wired as a *straight-through* cable, which means that the color of the wire on pin 1 at one end of the cable will be the same as pin 1 on the other end, pin 2 will be the same as pin 2, and so on. It will be wired to the T568A wiring standard contained in the TIA/EIA-568-B.1 standard, which states which color wire goes to each pin.

Patch cables are wired straight through because the cable from the workstation to the hub/switch is ordinarily crossed by the port on the hub/switch automatically. This is why many hubs/switches have an X next to their ports (meaning that the send and receive pairs will be crossed when the cabling connects to the hub/switch). The pinouts will be T568A, and all eight conductors (wires) should be terminated with RJ-45 modular connectors. (Only four of the eight wires are used for 10/100Base-T Ethernet; only two, four, or six of these wires will be used for telephony. All eight are used for 1000Base-T Ethernet.)

Patch cables must conform to the structured cabling standards and are considered to be part of "horizontal" cabling. This means that they are limited to 100 meters total between a workstation (such as a PC or printer) and a hub/switch. A patch cable can be used in a workstation area to connect the workstation network interface card (NIC) to an RJ-45 data jack in a wall outlet, or it can be used in the wiring closet to connect an RJ-45 jack on the patch panel to a port on a hub/switch. In addition, wires of this category and construction can be used for any form of residential telephony.

WHAT YOU WILL DO

▸ Build a straight-through Ethernet patch cable for use in a residential network.

▸ Test the patch cable for good connections (continuity) and correct pinouts (correct color of wire on the right pin).

▸ Demonstrate your understanding of the T568A wire scheme used by the Leviton line of home networking products.

▸ Inspect your wiring using appropriate tools.

WHAT YOU WILL NEED

▸ CAT 5, 5e, or 6 stranded cable

▸ Two RJ-45 plugs

▸ Cable jacket removal tool

▸ RJ-45 crimp-connector tool

▸ Cable cutting tool

Part 1: Preparation

Instructions are provided here for building a T568A type cable. A patch cable that is wired straight-through will have the same color of wire on the same pin (1–8) at both ends. Table 3-1-1 summarizes the T568A cabling standard:

Pin #	Pair #	Function (10Base-T)	Wire Color	Used with Telephony	Used with 10/100 Base-T Ethernet?	Used with 100Base-T4 and 1000Base-T Ethernet?
1	3	Transmit	White/Green	Line 3	Yes	Yes
2	3	Transmit	Green/White	Line 3	Yes	Yes
3	2	Receive	White/Orange	Line 2	Yes	Yes
4	1	Not used	Blue/White	Line 1	No	Yes
5	1	Not used	White/Blue	Line 1	No	Yes
6	2	Receive	Orange/White	Line 2	Yes	Yes
7	4	Not used	White/Brown	Line 4	No	Yes
8	4	Not used	Brown/White	Line 4	No	Yes

Table 3-1-1 T568A cabling

Part 2: Cutting and Stripping the Cable

1. Cut a six-inch piece of CAT 5e or greater unshielded twisted-pair cable. Figure 3-1-1 shows unshielded twisted-pair cable.

Figure 3-1-1 Unshielded twisted-pair

2. Use the stripping tool to remove approximately 3 inches of the outer jacket. Open the tool, push the cable through the proper opening, and turn the tool around the cable one turn. Open the tool and remove the cable. The outer jacket can now be pulled off. Figure 3-1-2 shows a cable stripper.

3. The tool is designed to score the jacket, not necessarily cut all of the way through. If the jacket cannot be pulled off, bend the cable gently from side to side at the point it was scored, then pull the jacket off.

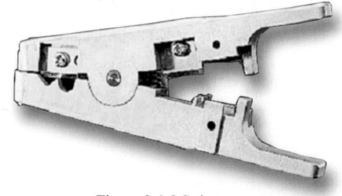

Figure 3-1-2 Stripper

Part 3: Organize the Wires

1. Hold the 4 pairs of twisted cables tightly where jacket was cut away, and then reorganize the cable pairs into the order of the T568A wiring standard:

   ▶ White/Green

   ▶ Green

   ▶ White/Orange

   ▶ Blue

   ▶ White/Blue

   ▶ Orange

   ▶ White/Brown

   ▶ Brown

2. Flatten, straighten, and line up the wires, then trim them in a straight line to within ½ inch from the edge of the jacket. Be sure not to let go of the jacket and the wires, which are now in order. You should minimize the length of untwisted wires because overly long sections that are near connectors are a primary source of electrical noise.

Part 4: Insert the Conductors

1. Place an RJ-45 plug on the end of the cable, with the tab on the underside and the green pair at the top of the connector. Figure 3-1-3 shows the RJ-45 plug.

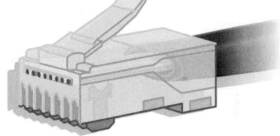

2. Gently push the plug onto the wires until you can see the copper ends of the wires through the end of the plug. Make sure that the end of the jacket is inside the plug and all wires are in the correct

Figure 3-1-3 RJ-45 plug

order. If the jacket is not inside the plug, it will not be properly strain relieved and will eventually cause problems.

Part 5: Crimping the Plug

1. If the wires are in the correct order, if they reach the end of the plug, and if the end of the jacket is inside the connector, use a ratcheting crimper to seat the pins by inserting the plug into the crimper and squeezing the handles

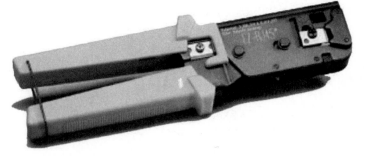

Figure 3-1-4 Crimping tool

together all of the way until the ratchet mechanism releases the handles. There is no need to press harder than the effort required to complete the ratchet cycle. Figure 3-1-4 shows the crimping tool.

2. Repeat Parts 3–5 to terminate the other end of the cable, using the same scheme to finish the straight-through cable.

Part 6: Inspecting the Cable

Two methods can be used to check the cable:

▶ **Visual Test.** Inspect the cable ends visually. Hold the RJ-45 connectors side by side; the same color wire should be on the same pin. This is not a conclusive test, but it is a good start.

▶ **Cable Test.** Use the LAN Pro Navigator to test your cable. Plug one end into the main unit and the other end into the remote unit. Turn the main unit on and press Test. If the cable does not pass the test, one or both ends need to be crimped again by cutting off the connector, restripping the cable, and following parts 3 through 6 again. Figure 3-1-5 shows the cable tester.

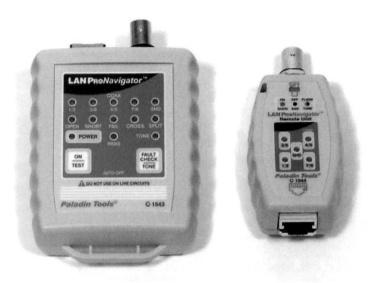

Figure 3-1-5 LAN Pro Navigator

Installing a Modem

INTRODUCTION

A modem modulates outgoing digital signals from a computer to analog signals for a telephone line, and demodulates the incoming analog signal and converts it to a digital signal for the computer. Modems come in two types: internal and external. Internal modems are installed inside a computer. External modems are connected via a serial communications port to the computer. Most new personal computers come equipped with 56K dial-up modems.

WHAT YOU WILL DO

In this lab, you will install a dial-up modem and then test it.

WHAT YOU WILL NEED

- ▶ PC
- ▶ Internal modem
- ▶ ESD grounding strap
- ▶ Screwdrivers

Part 1: Preparing the Computer

1. Make sure that the computer is unplugged. Place it on its side, open the case, and remove the cover.

2. Put on an electrostatic discharge (ESD) grounding strap and clip the other end to an unpainted metal surface of the computer case. This will place your body at the same electrical potential as the computer and its parts. See Figure 3-2-1.

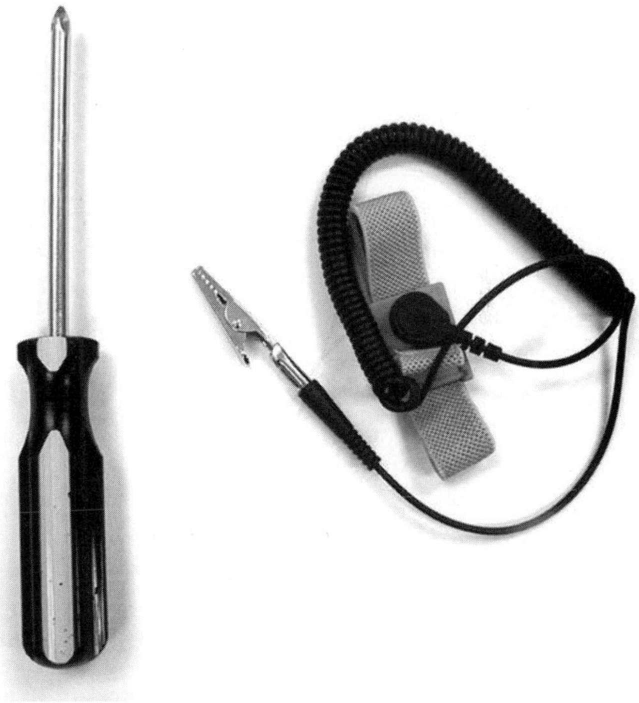

Part 2: Inserting the Modem

1. Remove the metal slot cover of an empty Peripheral Component Interconnect (PCI) slot.

Figure 3-2-1 ESD strap and screwdriver

Computer motherboards may have two or more types of slots: PCI and ISA are the most common. PCI slots are slots that are generally shorter and colored white. PCI slots are usually closest to the central processing unit (CPU) beside the Advanced Graphics Port (AGP) slot that is smaller (and there will only be one slot). See Figure 3-2-2.

2. Compare the card edge of the modem to the slot before attempting to install the modem into a slot.

3. Hold the modem by the metal carrier and top edge and insert it, pin-side down, into the adapter slot.

4. Push the card until it is in place. Never force the modem card into position or it may be damaged. You may need to rock the card gently to insert it.

Figure 3-2-2 PCI modem

5. When the card is properly seated, tighten the slot keeper screw to secure it into place.

6. Unhook the ESD strap from the computer and replace the cover carefully.

7. Plug in any device connectors that may have been removed to open the case.

Part 3: Verifying that the Modem Is Installed

Connect a phone cord to the modem and to a telephone outlet and switch on the PC. During the boot sequence, Windows should detect the new modem. If the modem is not detected, you can install it manually:

1. Select Windows Start, Settings, Control Panel, and then Modems. Click on Next and allow Windows to detect your modem. See Figure 3-2-3.

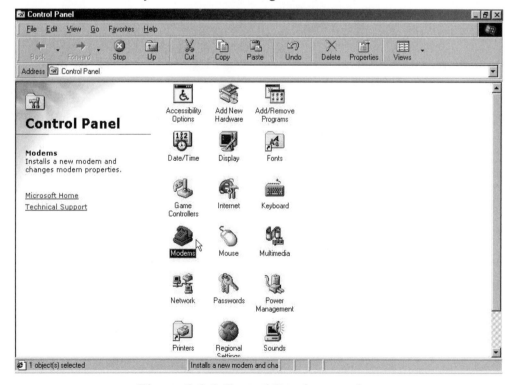

Figure 3-2-3 Control Panel screenshot

2. If Windows is still unable to detect the modem, select the modem from a list. Click on Have Disk and insert the manufacturer's driver disk and follow the prompts. If you do not have a disk, identify the modem from the list offered and click on Next.

Part 4: Installing the Driver

After the modem is detected, Windows will prompt you to insert the driver disk that came with the modem card.

1. Insert the disk and follow the installation directions.

2. After the driver is installed, you will need to select the COM port to assign it to.

Part 5: Performing a Self-Test on the Modem

1. Click on Start, Settings, Control Panel, and Phone and Modem Options.

2. Select the Diagnostics tab.

3. Highlight the installed modem and click on More Info.

The modem will do a self-test and display the results.

Part 6: Setting the Dialing Properties

1. Click on Start, Settings, Control Panel, and Phone and Modem Options.

2. Enter the area code and whether the phone is tone (most modern phones—you hear different pitches for each number pressed) or pulse (old phones—you hear eight clicks if you select the 8 button).

   Look at the Modems tab to see the modem that is installed.

3. Click on the Properties button. You will see the speed setting.

Conclusion
The modem is now installed.

Installing a Network Interface Card

INTRODUCTION

A network is defined as two or more computers linked together for the purpose of communicating, sharing information, and sharing other resources. In order for a PC to communicate on a network, three basic requirements must be met: The network must provide connections, communications, and services.

A NIC is a computer circuit board or card that is installed in a computer so that the PC can be connected to a network. The NIC provides an interface to the media (cabling, wireless) and enables communication with other PCs or devices. It translates the *parallel* signal produced by the computer into the *serial* format that is sent over the network. The installation is the same as for any other expansion card. After the NIC has been installed, it will then require configuration of the following system resources:

- ▶ IRQ (interrupt request)

- ▶ I/O address

- ▶ Memory address

Typically, a NIC uses IRQ 3 or IRQ 5. If the NIC is PnP (plug-and-play), these system resources will be configured automatically when the computer is powered on and the new hardware is detected.

WHAT YOU WILL DO

- ▶ Install a NIC.

- ▶ Test the NIC for connectivity.

- ▶ Configure a static IP address and subnet mask.

WHAT YOU WILL NEED

- ▶ PC

- ▶ Phillips screwdriver

- ▶ 10BaseT Ethernet patch cable

- ▶ 10BaseT NIC (PCI)

- ▶ Disk that comes with your NIC

- ▶ ESD grounding strap

Part 1: Preparing the Computer

1. Make sure that the computer is unplugged.

2. Place it on its side, open the case, and remove the cover.

3. Put on an ESD grounding strap and clip the other end to the computer case.

Part 2: Inserting the Network Card

1. Remove the metal slot cover of an empty PCI slot.

2. Hold the NIC by the metal carrier and sides and insert it, pin-side down, into the adapter slot.

3. Push the card until it is in place. Never force the network card into position or it may be damaged.

4. When the card is properly seated, tighten the slot cover screw to secure it into place.

5. Replace the cover and plug the power cord into the computer.

Part 3: Verifying that the NIC Is Installed

1. Turn on the computer and plug the Ethernet cable into the Ethernet port on the back of your NIC.

2. Plug the other end of the Ethernet cable into the DATA port on the faceplate in the wall.

During the boot sequence, Windows should detect the new network card. If the network card is not detected, you can install it manually:

1. Select Start, Settings, Control Panel, and then Add/Remove New Hardware.

2. Click on Next and allow Windows to try and detect your NIC.

If Windows is still unable to detect the NIC, do the following:

1. Select the NIC from a list.

2. Click on Have Disk, insert the manufacturer's driver disk, and follow the prompts. If you do not have a disk, identify the NIC from the list offered and click on Next.

Was your NIC card automatically detected on boot-up? _____

Part 4: Installing the Driver

A driver is a piece of software that allows the operating system to access a piece of hardware. After the NIC is detected, Windows will prompt you to insert the driver disk that came with it. Insert the disk and follow the installation directions.

Part 5: Verifying the NIC was Installed Properly

1. Right-click on the My Computer icon on the desktop.

   A shortcut menu will appear.

2. Click on Manage (see Figure 3-3-1).

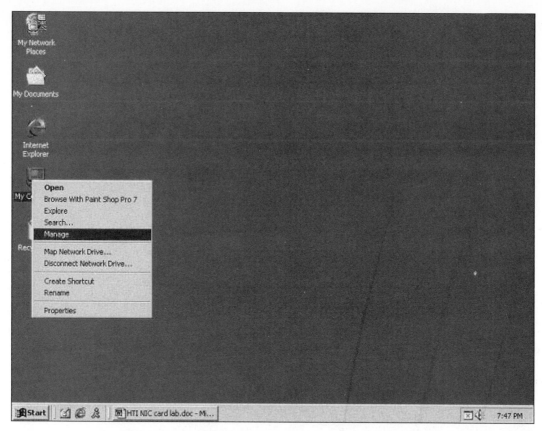

Figure 3-3-1 Click on Manage in the shortcut menu

3. Click on Device Manager to display the system's hardware devices (Figure 3-3-2).

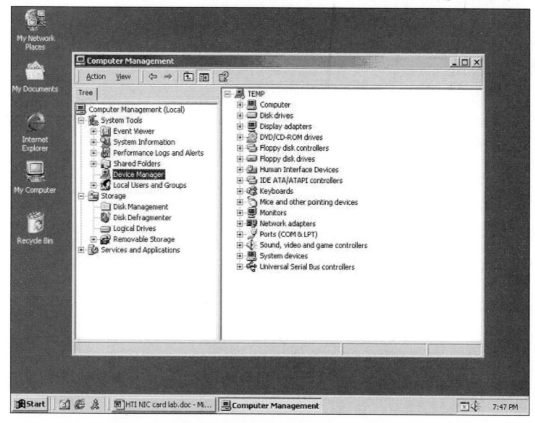

Figure 3-3-2 The system's hardware devices display

Installing a Network Interface Card   19

4. Click on the "+" sign next to "Network adapters" (see Figure 3-3-3).

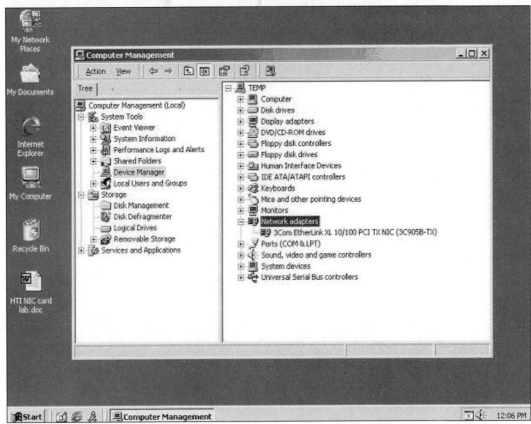

Figure 3-3-3 Displaying network adapters

Do you see the name of your NIC here? _____

What type of NIC is installed in your computer? _____

_____

5. Double-click on the type of NIC installed in your computer.

6. Select the Resources tab to identify the system resources configured for this adapter card (see Figure 3-3-4).

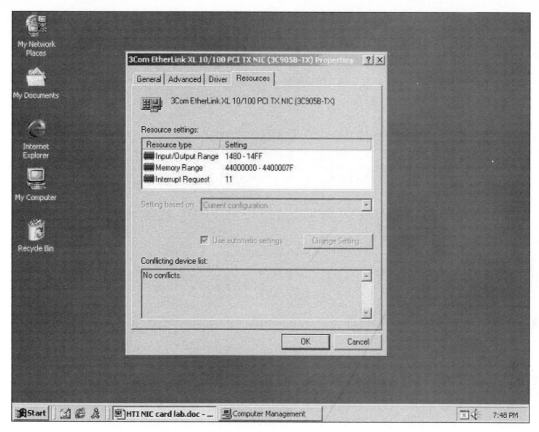

Figure 3-3-4 Identifying system resources

What IRQ is being used? _____

What is the I/O Address? _____

What is the Memory Range? _____

Part 6: Configuring a Static TCP/IP Address

1. Right-click on the My Network Places icon on the desktop. A shortcut menu will appear.

2. Click on Properties (see Figure 3-3-5).

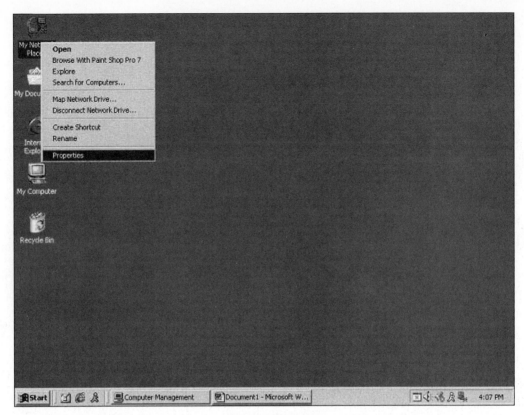

Figure 3-3-5 Click on Properties in the shortcut menu

The Network Dial-up Connection window will open.

3. Right-click on Local Area Connection and click on Properties (see Figure 3-3-6).

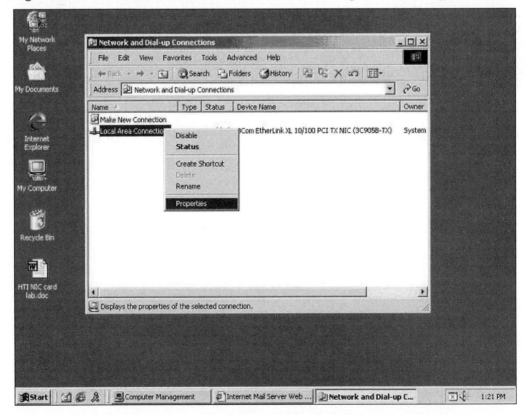

Figure 3-3-6 Click on Properties under Local Area Connection

4. Click on Internet Protocol (TCP/IP)

5. Select Properties.

   This will bring up your Internet Protocol (TCP/IP) Properties window (see Figure 3-3-7).

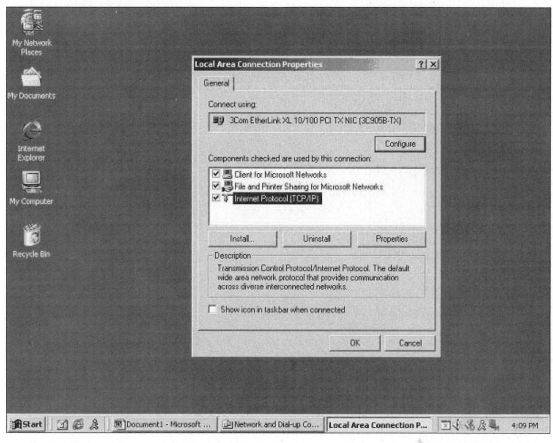

Figure 3-3-7 The Internet Protocol (TCP/IP) Properties window displays

6. Click on the radio buttons Use the following IP address and Use the following DNS server addresses.

7. Enter the information that your instructor has given you in the blank form that follows the next screen display. If your classroom is using DHCP, however, you should select the following radio buttons instead: Obtain an IP address automatically and Obtain DNS server automatically (see Figure 3-3-8).

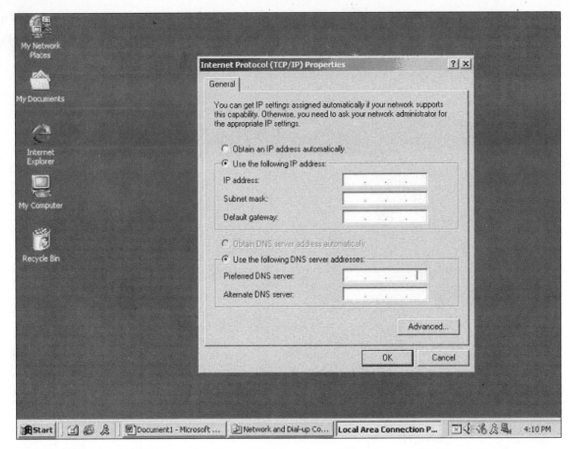

Figure 3-3-8 Enter the IP and DNS server addresses on the General tab

IP Address	
Subnet Mask	
Default Gateway (optional)	
Preferred DNS server (optional)	

8. When you have finished entering your IP address, Subnet Mask, Default Gateway (optional), and Preferred DNS server (optional), click on OK.

Part 7: Testing for Connectivity Using the Ping Command

The ping (Packet Internet Groper) command is a useful command-line utility that sends a message called an Echo Request, by using Internet Control Message Protocol (ICMP), to a destination computer and tests for reachability of a network device. The destination computer responds by sending an ICMP echo reply. Ping can be used with either the hostname or an IP address to test IP connectivity. If you receive a reply, you know that the physical connection between the two devices is intact and working.

1. Click on the Start button.

2. Select Run.

3. Enter **cmd** (or enter **command** if using Windows 95, 98, or Me).

4. Press Enter.

5. At the command prompt, enter **ping 127.0.0.1**.

6. Press Enter.

Does your screen look like this? (See Figure 3-3-9.) _____

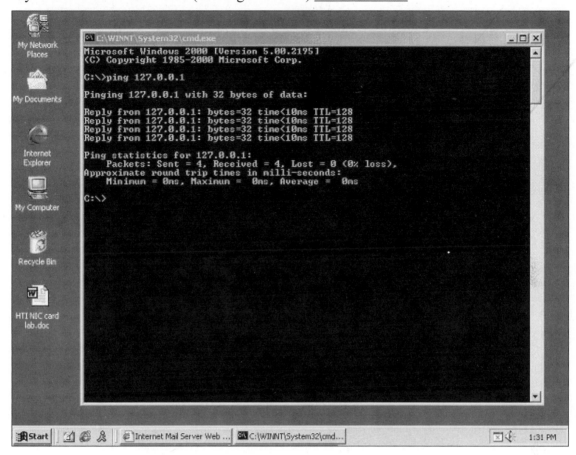

Figure 3-3-9 Your screen should look like this.

Networking PCs

INTRODUCTION

Computers are networked to share resources. Networking is a physical connection between computers and/or network devices. Two computers can have a single cable between them. Connecting multiple computers requires an interconnection device, which can be either a hub or a switch.

In addition to a physical connection, computers must be able to identify themselves on the network, and they must speak the same language. Ethernet is the language these computers will speak, and TCP/IP is the way that computers use to identify themselves. TCP/IP is called a protocol. There are other protocols, such as IPX and NetBEUI, but TCP/IP is the most popular protocol used on the Internet.

WHAT YOU WILL DO

▶ In this lab, you learn how to connect two PCs to create a basic network. You will then configure TCP/IP protocol settings for the two PCs.

▶ Finally, you will verify that the PCs can communicate by using the ping command.

WHAT YOU WILL NEED

▶ Two PCs with network interface cards (NICs) preinstalled

▶ Hub

▶ Straight-through Category 5 cable (from a previous lab)

Setting Up

This lab uses common operating systems found in residential networking environments, such as Windows 95 or 98. To create the network, two PCs will be connected to a hub via Category 5 unshielded twisted-pair (CAT 5 UTP) straight-through cables with RJ-45 connectors. Hubs can be used to connect multiple PCs and other equipment, such as printers and scanners.

Part 1: Setting Up the Hub

Place the hub between the two PCs, and plug the hub's AC adapter into a power outlet. Figure 3-4-1 shows a hub.

Figure 3-4-1 Hub

Part 2: Connecting the CAT 5 Cable

1. Locate the RJ-45 ports on the network interface cards (NICs) of the PCs.

2. Connect one end of the straight-through CAT 5 cable to the first PC and run the other end into an available port on the hub.

3. Do the same for the other PC. See Figure 3-4-2.

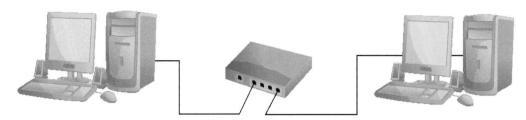

Figure 3-4-2 Connecting PCs

Part 3: Checking Connectivity

Turn on the PCs. The link light on the hub verifies that the PC NIC and the hub port have a good connection. If the link light is not lit, it may be that the plugs are not inserted correctly, the cable is incorrectly wired, or one of the devices is not working properly.

Part 4: Verifying the NIC

On each PC, verify that the NIC is functioning properly.

1. From the Start menu, choose Settings, Control Panel, System, and Device Manager.

2. Double-click on Network Adapters and then right-click on the NIC adapter in use.

3. Click on Properties to see whether the device is working properly. If there is a problem with the NIC or driver, there are two possible reasons:

   ▶ The icon will show a yellow circle with an exclamation mark in it, indicating possible resource conflict.

   ▶ The icon will show a red X, indicating a serious problem that could cause Windows to lock up.

If there is any problem with the NIC in the Device Manager, ask your instructor to help you resolve the issue.

Part 5: Checking the TCP/IP Protocol Settings

You will need to check the TCP/IP protocol settings on both workstations.

1. Select Start, Settings, Control Panel and then choose Network.

2. Choose the TCP/IP protocol from the Configuration tab and then click on the Properties button.

3. Check the IP Address and Subnet Mask on the IP Address tab. Figure 3-4-3 shows a screen shot of the TCP/IP Properties window.

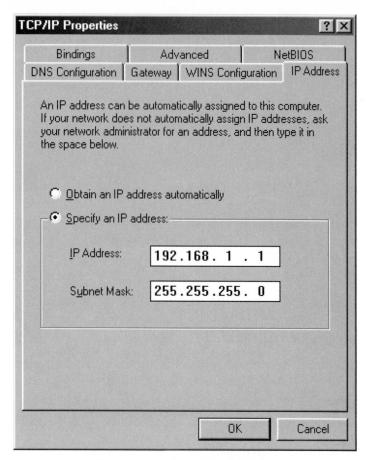

Figure 3-4-3 TCP/IP properties

An IP address is a logical address that is assigned to a computer.

For this lab, we will use the private IP address of 192.168.1.0. Private IP addresses are for internal use on a home or business network and cannot be used on the Internet.

1. Configure the first workstation to have the statically assigned IP address of 192.168.1.1 and the second to have the IP address of 192.168.1.2 for our local area network.

2. Configure the subnet mask on both workstations to be 255.255.255.0.

Network devices such as PCs use subnet masks to determine whether they are on the same or a different network as another device so that communication can be established and traffic across networks can be minimized.

After you have configured the IP addresses, Windows will prompt you to reboot the workstation.

Part 6: Verifying TCP/IP Settings
Windows 95, 98, or Me:

1. From the Start menu, choose Run. Type **winipcfg** in the window and click on OK. This command lets you see the TCP/IP settings.

   The IP Configuration dialog box appears. The adapter address is the NIC address, which is also known as the Media Access Control (MAC) or physical address. Unlike the IP address, the MAC address does not change; it is a number assigned by the manufacturer to the NIC. If the NIC is replaced, the physical address changes to the address of the new NIC.

2. Locate the section in the IP Configuration dialog box that displays the Ethernet Adapter (NIC) being used. If the Ethernet Adapter is not already displayed click on the down arrow at the right of the display box and scroll down to select it.

3. Close the window by clicking on OK.

Windows 2000 or XP:

1. From the Start menu, choose Run.

2. Type **cmd** in the window and click on OK.

3. After the command line window opens, type **ipconfig/all** and press Enter. This command lets you see the TCP/IP settings.

4. When you are finished, the window can be closed by typing **exit** and pressing Enter.

Part 7: Pinging the PCs

Use the ping command to check for basic TCP/IP connectivity.

1. Click on Start, Programs, Accessories and then select the MS-DOS prompt or Command prompt, depending on the operating system installed.

2. Type **ping** followed by the IP address of the other PC (**C:\>ping 192.168.1.2**) and then press Enter. Figure 3-4-4 shows the command line.

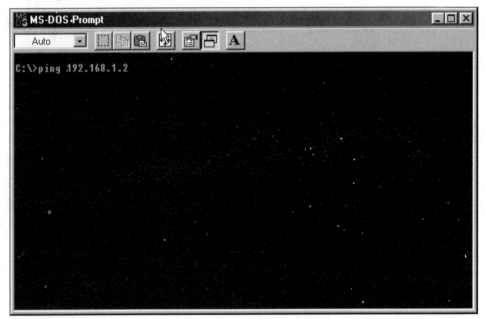

Figure 3-4-4 Command line

You will see the message "Reply from 192.168.1.2. Ping successful" or "Request timed out." The request timed out message indicates that there is no network connection between the two PCs.

Part 8: Troubleshooting

If you saw the message "Request timed out," you will need to troubleshoot the problem:

1. Check the link light of the port that the PC is connected to on the hub. If the link light is not lit, check the cable connection.

2. If the link light is lit, confirm the address of the PC that you are attempting to ping by repeating Part 6. Make any necessary changes and repeat Parts 6, 7, and 8.

Configuring the Leviton Residential Gateway

INTRODUCTION

A big part of home integration technology revolves around the capability to efficiently move data between devices within the home. As the Internet has become integrated into our lives, there has developed a need to exchange information with the outside world for access to the World Wide Web and e-mail. This integration also has some obstacles. For example, most Internet service providers (ISPs) provide a single address for you to access the Internet unless you want to pay for more. So the first problem is to connect multiple computers simultaneously to the Internet by using a single connection. Another problem is keeping personal data private while sharing certain data with the outside world.

An Internet gateway provides a solution to both of these problems. The problems outlined previously are solved by two functions that the gateway provides: Network Address Translation (NAT) and Dynamic Host Configuration Protocol (DHCP).

NAT allows a connection to the Internet through a single shared connection provided by the ISP. The key is in the implementation of public and private IP addresses. Every valid Internet connection must have a valid public IP address; public IP addresses are required for each computer connected to the Internet. This address is used to route network messages properly from one location to another. Private addresses, on the other hand, are designed for personal use and therefore are not routable. A NAT server enables you to share a single connection to the Internet by using a single public address to send out requests on the Internet on behalf of computers with addresses from the private range. NAT translates the messages from the privately addressed computers to the Internet. Messages coming in from the Internet are all sent to the public IP address assigned to the gateway and translated back to the appropriate computer using its private IP address.

The next service that the Leviton residential gateway provides is Dynamic Host Configuration Protocol (DHCP), which automatically assigns an IP address within the private IP address range to the computers on your network. This allows you to configure up to 254 computers automatically in order to share a single Internet connection.

WHAT YOU WILL DO

- ▸ Install the Leviton residential gateway into the Structured Media Center (SMC).
- ▸ Wire the Leviton residential gateway to provide Internet access to multiple PCs.
- ▸ Test a "live" connection to the Internet using the connectivity lights.

WHAT YOU WILL NEED

- ▸ Leviton residential gateway
- ▸ PC with network interface card and web browser
- ▸ Two patch cords with RJ-45 connectors
- ▸ Leviton voice and data module

▶ Operational Internet connection in the classroom

OPTIONAL ITEMS

▶ URL: http://www.leviton.com/sections/pr/articles/Providingbroad.pdf

▶ URL: http://www.homenethelp.com/router-guide/index.asp

SETUP

The instructor will run a Category 5e cable between the SMC and a CAT5 jack. Make sure that the jack is wired to the 568A wiring scheme. Terminate the end of the cable in the SMC to position 3 of a voice and data module. This jack will simulate a location within the home that requires Internet access.

Part 1: Attaching the Bracket to the SMC

Secure and lock the gateway into place in the Structured Media Center (SMC) with push lock pins:

1. Align the tabs on the mounting bracket with the holes on the SMC and hook them into place. (See Figure 3-5-1.)

2. Ensure that the push lock pins are in the "open" position, with the pins pulled partially out of the bases, but not removed.

3. Press the push lock pins all the way down to lock them into place.

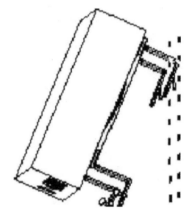

Figure 3-5-1 Attaching the gateway on the SMC

Part 2: Cabling the Internet Gateway

The Internet gateway has five connection ports. Four ports can be used to connect a variety of equipment including computers, printers, scanners, and other network peripherals. The fifth port is for a WAN connection. This is the connection to the DSL or cable modem (or a switch or router, depending on the size of the network).

1. At the SMC, power the gateway using the supplied AC adapter.

2. At the SMC, locate the Internet gateway port marked WAN. Connect this port to an Internet-capable jack in the classroom. This connection will supply the gateway with Internet connectivity.

3. Be sure that the Uplink button on the back of the gateway is pushed in.

This simulates the connection one would have from a cable or DSL modem.

What light on the gateway is lit? _____

This is the Link light. This light shows that data is reaching the gateway from the Internet connection.

1. Connect a patch cord between one of the gateway's remaining ports and the third port from the top on the voice and data module.

2. Connect a patch cord between the CAT 5e jack on the lab wall to which the voice and data module is connected and the network card in the PC.

Are any lights blinking on the network card? _____

What lights are lit or blinking on the gateway?

_____

_____

What number is indicated underneath the blinking lights on the gateway? _____

Does this number correspond with the number indicated on the back of the gateway where the connection to the jack is plugged in? _____

Draw a small sketch or schematic of the cabling and components.

This simple observation of blinking lights on the gateway and on the network card is the first indication that there is connectivity throughout the cabling from the PC through the jack and cabling, and through the voice and data module, gateway, and cable or DSL modem. If there are any connectivity issues, check all of the connections and trace the route of the wires from the PC to the Internet connection in the classroom.

Part 3: Accessing the Gateway Configuration

The gateway has a web-enabled interface. That means you access it with a browser, just like accessing pages on the Internet. The default IP address of the gateway is 192.168.1.1.

1. Double-click on the Internet Explorer icon on the desktop.

2. Enter **http://192.168.1.1** into the URL field of the browser window (see Figure 3-5-2)

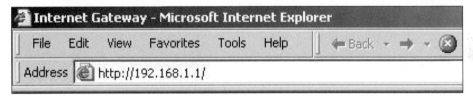

Figure 3-5-2 IP address

3. Accessing the gateway in this fashion allows you to configure the gateway's settings.

4. After the gateway is contacted, you should see a prompt for a user name (administrator name) and password. Enter the word *root* as the administrator name, and leave the password field empty (see Figure 3-5-3).

Figure 3-5-3 Enter password

5. Click on OK. Figure 3-5-4 show the start page for the gateway interface.

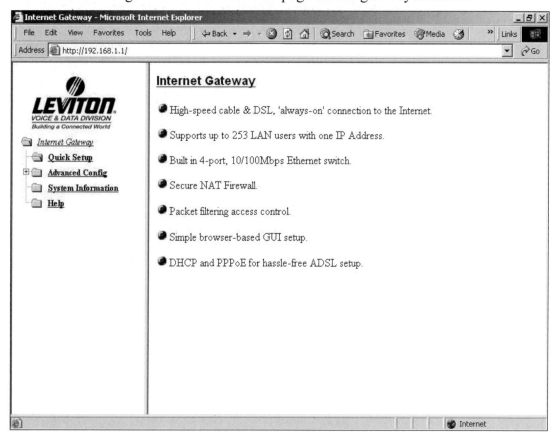

Figure 3-5-4 Gateway interface

6. Select Quick Setup (on the left) from the next window. You will be routed to the Quick Setup window.

7. Use the Quick Setup window to specify how your 10/100 Mbps Internet gateway will obtain an IP address (Dynamic, Fixed, or PPPoE). Choose Dynamic.

Part 4: Dynamic Addressing (DHCP)

The default setting causes the gateway to receive its address dynamically (automatically) from the ISP server. Enter any relevant information and click on Apply (see Figure 3-5-5).

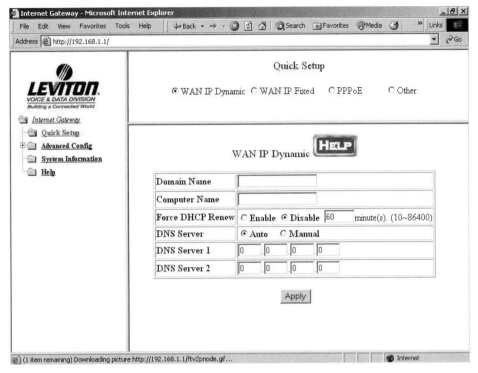

Figure 3-5-5 Dynamic addressing window

Part 5: Fixed IP Address

If your ISP has provided a fixed IP address, click on the WAN IP Fixed selection near the top of the window. You will be routed to the Quick Setup—WAN IP Fixed window (see Figure 3-5-6).

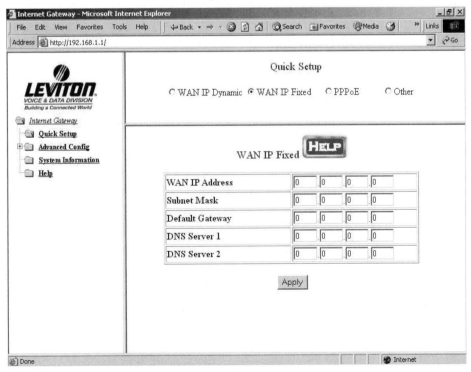

Figure 3-5-6 Fixed IP address window

Configuring the Leviton Residential Gateway   35

Enter the appropriate information, as provided to you by your Internet Service Provider, click on Apply, and then skip to Part 7.

Part 6: PPPoE Address

1. If your ISP supports PPPoE (Point-to-Point Protocol over Ethernet), click on the PPPoE selection near the top of the window. You will be routed to the Quick Setup—PPPoE window (see Figure 3-5-7).

Figure 3-5-7 PPPoE address window

2. Enter the appropriate information, as provided to you by your Internet service provider and then click on Apply.

Part 7: Finishing Up

1. Record your IP address for future reference.

2. After you click on Apply, the prompt will appear. Click on OK to reboot your gateway and activate the new configuration information (see Figure 3-5-8).

Figure 3-5-8 Saving settings

The Internet gateway is now ready for use. All the network devices you connected will now be able to access Internet services simultaneously through the 10/100 Mbps Internet gateway.

Setting Up a Firewall

INTRODUCTION

When computers were first networked, there was little need for security. The inventors of the Internet were interested in harnessing the power of networking computers to share research information. During this time, malicious activity was not an issue. Because Internet popularity has exploded with attractions such as online shopping and banking, the world has become ever more dependent upon its use. Certain individuals have recognized the ability to profit from it in many different ways and have been working on methods to exploit it. These exploiters, often called hackers, crackers, or black hats, have many reasons for doing what they do. Some do it for financial gain, espionage, personal satisfaction, and even fun.

When hacking was new, it took experienced, well-trained individuals with expensive equipment and a lot of time to compromise a computer system. However, with the recent widespread availability of inexpensive, high-speed Internet access coupled with a significant increase of hacker-related resources, even a rank amateur can download hacking tools and use them on unsuspecting victims without even knowing how the underlying technology works.

So whether it is a company attempting to keep its secrets safe, or a home user wanting to harness the power of the Internet while keeping their personal information safe, it is necessary to employ protection known as a firewall. Firewall products come in a wide variety of shapes, sizes, and costs. Some are hardware-based; some are software-based; some are a combination of hardware and software often called firmware.

So, what type of firewall is right for you? That depends on what you are protecting, and (of course) your budget. For a company such as eBay, which operates solely on the Internet and needs to protect not only its own assets but also the private information of its customers, an elaborate (and expensive) solution is used. For the home Internet user, a bit less drastic measure, which is less expensive and less complicated to configure, would probably suffice. For this reason, many home Internet users use software-based firewalls. One such software firewall, available from a company called Zone Labs, Inc., is called ZoneAlarm. ZoneAlarm is popular because it is effective, it is easy to install and configure, it is free for certain users, and it is inexpensive for all others.

WHAT YOU WILL DO

▸ Download and install a common software firewall product called ZoneAlarm (Figure 3-6-1).

▸ Configure the ZoneAlarm software firewall product to deny outside access to your PC.

▸ Perform tests to verify proper configuration and behavior of the ZoneAlarm software firewall.

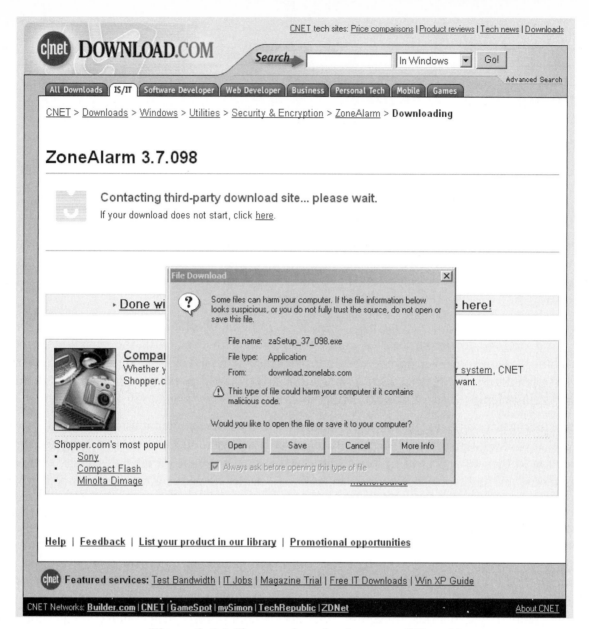

Figure 3-6-1 Shows a download screen of ZoneAlarm

WHAT YOU WILL NEED

▶ Two PCs with ZoneAlarm-compatible operating systems such as Windows 98, NT, 2000, or XP

▶ Proper network infrastructure to enable communication between the two PCs

▶ Internet access or a local copy of the ZoneAlarm firewall software installation files

OPTIONAL ITEMS

▶ http://www.firewall.com

▶ http://www.cert.org/tech_tips/home_networks.html

Setting Up

Make sure that you have two PCs configured with ZoneAlarm-compatible operating systems such as Windows 98. Set up a share on the PC on which you will be installing the ZoneAlarm product so that you will be able to test your configuration (or both PCs for more advanced testing). Make certain that the two PCs can communicate by using the ping command or by browsing using Network Neighborhood. Please refer to previous lab exercises or see your instructor for additional assistance.

Part 1: Installing the ZoneAlarm Software Firewall

Note: Depending on the version of ZoneAlarm you download, the installation may vary slightly. Please ask your instructor if you need help.

1. Download the latest free version of ZoneAlarm from www.zonelabs.com, if necessary.

2. Double-click on the installation program to begin. The ZoneAlarm installation begins with a warning to exit all Windows programs and prompts you for the desired installation location.

3. Select the default location by clicking on the Next button.

4. You will be prompted to enter user information. Complete the form, making sure to leave the registration and Inform me about updates check boxes empty.

5. Click on the Next button to continue.

6. Read and accept the License Agreement, check the box accepting the terms of the License Agreement, and then click on the Install button to continue. This will copy the necessary files to the PC.

7. You are prompted with a User Survey. Complete it; then click on the Finish button.

8. You will get a ZoneAlarm Setup box, confirming that the setup process is complete and asking you if you want to start the ZoneAlarm product. Click on the No button.

9. Reboot the PC.

Part 2: Configuring the ZoneAlarm Software Firewall

After your PC reboots, the ZoneAlarm Welcome screen appears. At this point, your firewall is not configured, so you need to do the following:

1. Click through the next few screens using the default settings. This will configure your Firewall with generally acceptable settings.

2. The last screen you come to, entitled Set-up complete, will launch the ZoneAlarm tutorial. Spend a few minutes going through the tutorial to familiarize yourself with the product and its various settings.

3. Secure your PC as much as possible using the information learned from the tutorial. If you get stuck, you can launch the Help menu in the top right of the ZoneAlarm window or ask your instructor for assistance.

Part 3: Testing your ZoneAlarm software firewall

1. Attempt to ping your ZoneAlarm-protected PC from another PC on the network.

   Was the ping attempt successful? _____

2. Attempt to ping another PC on the network from your ZoneAlarm-protected PC.

   Was the ping attempt successful? _____

3. Attempt to connect to your ZoneAlarm-protected PC from another PC on the network.

   Was the attempt successful? _____

4. Attempt to connect to another PC on the network from your ZoneAlarm-protected PC.

   Was the attempt successful? _____

5. Experiment with changing ZoneAlarm settings and testing your new configuration settings. Discuss your results with your classmates.

Conclusions

In today's Internet age, we rely on the ability to share information. The key to successful Internet use is to share the information intended, while simultaneously keeping other information private. With the widespread desire and ability of hackers to exploit our privacy, it is becoming ever more important to protect ourselves. You have seen that implementing a personal software firewall is a step in the right direction.

Configuring Ports

INTRODUCTION

We use computers to share all types of information. As computers become more integrated into our lives, we tend to ask more and more of them. We use them to surf the Internet, share data, and exchange e-mail for our personal and business lives. For computers to satisfy our requirements to do more and more, it has become necessary to implement standards to allow them to exchange information and perform various tasks. These tasks are handled by running small programs on our computers that we call services. Common services allow us to access web pages, transfer files, and exchange e-mail, to name a few. With a single computer running so many services, it is necessary to enable the computer to sort and route messages to the appropriate service. An addressing scheme soon became mandatory.

Data is routed from computer to computer on the Internet by using a software-based IP address, which can be loosely compared with a U.S. postal mailing address. Because every PC that is directly connected to the Internet must have a unique IP address, messages will get to the appropriate machine. The problem is the routing of messages after they get to the appropriate computer. Consider for a moment a comparison of tenants in an apartment complex. Each tenant is a member of the same complex and therefore shares a mailing address. So how would messages sent to the same mailing address get to the correct tenant? Assigning an additional unique number to each tenant called a unit or apartment number solves the problem. When messages come into the complex, they are routed to the appropriate tenant using the unique apartment number. This procedure is similar to the process of message handling inside the PC. Locations of services inside a PC are addressed by using an object called a software port. Unlike some of the physical ports you've seen, such as RJ-45 and RJ-11s, these ports are logical locations in the software of the PC. A port is actually made up of two components: the IP address and a unique location in memory represented by a number. Like our message to the apartment tenant, the main address gets the message to the correct general area, and the port number gets the message to the appropriate service.

For standardization purposes, certain software ports have become industry standards: These are called well-known ports. Table 3-7-1 illustrates some common well-known ports in widespread use today (it is not an extensive list of well-known ports).

Network Service	Port Number
WWW (World Wide Web)	80
SMTP (Simple Mail Transfer Protocol)	25
FTP	Two ports = 20 used for data, 21 used for control
Telnet	23
POP3 (Post Office Protocol)	110
DNS (Domain Name System)	53

Table 3-7-1 Common ports

This arrangement works well for proper configuration and hosting of multiple services on a single PC; however, there is a drawback. Because all computers are using the same ports for common services, hackers have exploited this information to find vulnerabilities in your computer. To identify which services your PC is running, hackers generally employ a tactic called a port scan, which contacts certain well-known ports to see whether services are running and then attempts to exploit vulnerabilities in a particular service. For this reason, it is always a good idea to disable unused services on your PC and to employ additional technology such as a firewall to protect it against common types of network-based attacks. This involves blocking these ports on the firewall. Additional methods of self-protection involve assigning alternate port numbers to certain network services. Keep in mind, however, that non-customized PCs will ask for services using the well-known port number. Therefore, you must be careful to make configuration changes to all machines on your local network. Additionally, you may employ a technique called port forwarding, in which a residential firewall or gateway maps a well-known port to your customized location automatically.

WHAT YOU WILL DO

▶ Use your knowledge of IP addressing, modems, gateways, firewalls, and ports to identify specific ports that must be available to enable use of specific network services.

▶ Identify proper port forwarding settings to allow use of network services.

WHAT YOU WILL NEED

▶ This lab

▶ Pen or pencil

OPTIONAL ITEMS

URL: http://www.iana.org/assignments/port-numbers

Setting Up

Review the introduction to this exercise and previous labs relating to gateways, modems, and firewalls.

Part 1: Allowing Specific Ports Into your Network

1. Review the home network diagram shown in Figure 3-7-1. Use your knowledge of IP addressing, gateways, firewalls, and ports to determine specific ports that must be allowed into your home network through the gateway in order for your network services to operate properly.
2. Answer the questions in Part 3 regarding gateways and port forwarding.

Part 2: Port Access

1. Review the home network diagram shown in Figure 3-7-1.
2. Use your knowledge of IP addressing, gateways, firewalls, and ports to determine specific ports that must be allowed into your home network through the firewall installed on each machine in order for your network services to operate properly.

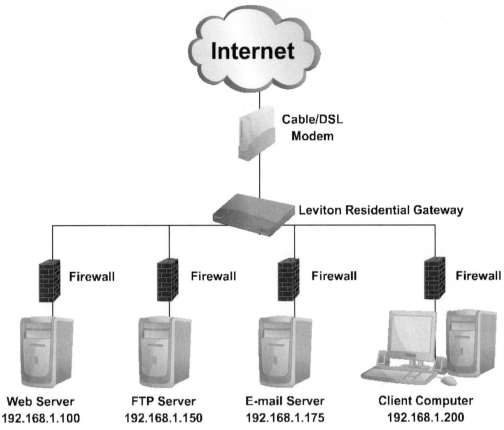

Figure 3-7-1 Home network diagram

Part 3: The Gateway

Port forwarding directs computers outside the network to computers inside the network that offer the services requested by computers on the outside. The computers inside the network are "listening" on designated ports.

1. Which ports must be forwarded to the computer with the address of 192.168.1.150?

    _____

2. Which ports must be forwarded to the computer with the address of 192.168.1.175?

    _____

3. Which ports must be forwarded to the computer with the address of 192.168.1.33?

    _____

4. Which ports must be forwarded to the computer with the address of 192.168.1.100?

    _____

Part 4: The Firewalls

Each computer will need to have a firewall installed to prevent malicious users from compromising the integrity of the system.

1. Which ports does the web server need to have open?

   _____

2. Which ports does the FTP server need to have open?

   _____

3. Which ports does the client computer need to have open?

   _____

Conclusions

A solid understanding of IP addressing, modems, gateways, and ports is mandatory to enable a properly functioning home network. Understanding these networking concepts and proper configuration techniques will enable you to host a variety of network services on your home network.

Home Networking Using NetBEUI

INTRODUCTION

Since the introduction of computers, their intrusion into our business and personal lives has been both extensive and extremely rapid. They enable us to share resources and enormous amounts of information like never before. However, their rapid evolution, coupled with many different ideas about the best way to connect them, has spawned a need for standards to exchange information.

Standards or rules for exchanging data between computers are called *protocols*. Protocols dictate every aspect of data transfer conversation between computers, including syntax, formatting, error correction, transfer rates, and addressing—just to name a few. For this reason, protocols are often compared to human languages such as English, French, and German. Like humans, computers must share a common protocol or language to effectively communicate.

One such protocol, implemented by IBM and then adopted and updated by Microsoft and Novell, is called NetBEUI. NetBEUI (pronounced net-BOO-eee) is an acronym for NetBIOS Extended User Interface. NetBEUI is a fast, efficient, and popular protocol that enables computers to share network resources such as files and printers.

Although many protocols are available to the home network user, the NetBEUI protocol does have some advantages over other networking protocols. First, it is widely known, accepted, and available. It is easy to set up and configure, even for users with very little knowledge of networking concepts. In addition, NetBEUI does not contain routing capabilities, and therefore is not routable to other networks without the assistance of another protocol. This benefits the home network user because it keeps NetBEUI network traffic local to your home network. This combination makes NetBEUI a good choice for home networking. The following lab demonstrates the ease of sharing data between computers using the NetBEUI protocol.

WHAT YOU WILL DO

▶ Learn how to connect two PCs together in a simple network and enable them to share data using the NetBEUI protocol. This lab example uses Windows 98, a common operating system found in residential networking environments. The two PCs will be connected to a hub using Category 5 unshielded twisted-pair (CAT 5 UTP) straight-through cables with RJ-45 connectors.

▶ Make the physical connections necessary to facilitate networking.

▶ Install and configure the NetBEUI protocol on both PCs.

▶ Share files on one of the PCs to allow data access over the network.

▶ Test connectivity by transferring data between the two PCs.

WHAT YOU WILL NEED

▶ Two PCs equipped with Network Interface Cards (NICs) and a NetBEUI-capable operating system such as Windows 98

▶ Windows 98 operating system disks

- One network hub with RJ-45 connection ports
- Two straight-through Category 5 unshielded twisted-pair patch cables

OPTIONAL ITEMS

- Additional PCs to add to the simple peer-to-peer network
- Additional network-enabled devices such as printers and scanners

Setting Up
Gather the appropriate hardware and software needed for the project.

Part 1: Making the Physical Network Connections

1. Place the hub between the two PCs and connect the power.

2. Connect the PCs' network interface cards (NICs) to the hub using the Category 5 UTP cables.

3. Make sure not to plug either NIC into the UPLINK port on the hub.

4. Power up the PCs and verify your physical connections by checking for illuminated link lights on the NICs and on the hub. See Figure 3-8-1.

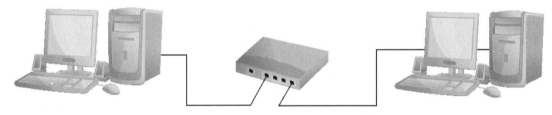

Figure 3-8-1 Connecting PCs

Part 2: Verifying the NIC
On each PC, verify that the NIC is functioning properly.

1. From the Start menu, choose Settings, Control Panel, System, and Device Manager.

2. Double-click on Network Adapters and then right-click on the NIC adapter in use.

3. Click on Properties to see whether the device is working properly.

If there is a problem with the NIC or driver, there are two possible reasons:

- The icon shows a yellow circle with an exclamation mark in it, indicating possible resource conflict.

- The icon shows a red X, indicating a serious problem that could cause Windows to lock up.

For more troubleshooting information, see Lab 3-3, "Installing a Network Interface Card," or ask your instructor for assistance.

Part 3: Installing the NetBEUI Protocol

1. From the Start menu, choose Settings, Control Panel, Network.

2. From the Configuration Tab, click on Add.

3. From the Select Network Component Type box, highlight Protocol and then click on Add.

4. From the Select Network Protocol box, highlight Microsoft in the Manufacturers section.

The Network Protocols section will change to display Microsoft-related network protocols.

5. Scroll down and highlight NetBEUI; then click on OK. See Figure 3-8-2.

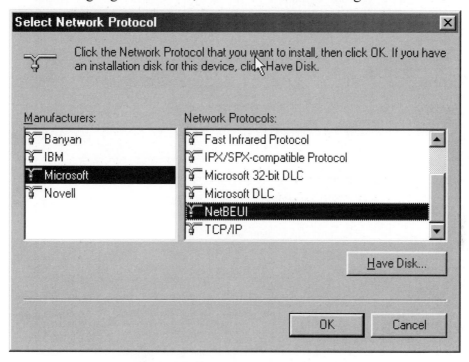

Figure 3-8-2 Screenshot of NetBeui

6. Insert the operating system CD if necessary.

7. Select Yes to reboot the system.

8. When the PC reboots, choose Settings, Control Panel, and Network from the Start menu.

9. From the Configuration tab, verify that the NetBEUI protocol has been installed.

10. Repeat step 3 on each of the PCs that will be used in the network.

Part 4: Configuring Systems for File Sharing

1. From the Start menu, choose Settings, Control Panel, and Network.

2. From the Network window, select the Identification tab.

3. Verify that you have a unique Computer Name for each PC and that *both* PCs have identical entries of HTI in the Workgroup field. *(Note:* You won't be able to complete this lab successfully if both PCs do not have identical Workgroup names, so watch your spelling!)

4. Optionally, enter a description of your choice in the Computer Description field.

5. When finished, click on OK. Verify this on each PC you are networking. If prompted for a reboot, select No.

6. From the Start menu, choose Settings, Control Panel, Network.

7. From the Configuration tab, click on the File and Print Sharing button.

8. From the File and Print Sharing box, check the entry I want to be able to give others access to my files; then click on OK (see Figure 3-8-3). Insert the Operating System CD, if prompted.

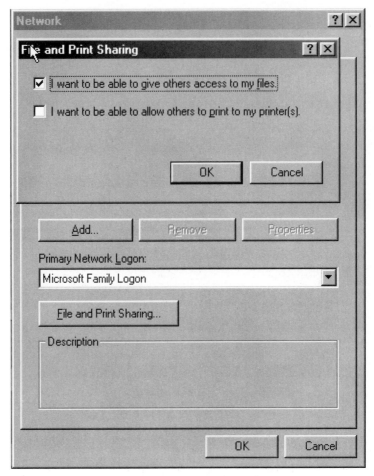

Figure 3-8-3 File and print sharing

9. Click on OK from the Network dialog box and allow the machine to reboot.

Part 5: Sharing Some Files
After the PCs are finished rebooting, configure a shared directory on one of the PCs.

1. From an empty area on the desktop, right-click and select New, Folder. Name the folder **Shared Docs**.

2. Enter the Shared Docs folder you just created by double-clicking on it.

3. Once inside, right-click in the white space and select New, Text Document.

4. Name the document **shared_document.txt** so you can read it later.

5. Double-click on the document icon to open it for editing.

6. Enter some text into the document and close it by clicking on the X in the upper-right corner of the window.

7. Answer Yes to the Save Changes prompt.

8. Close the Shared Docs folder by clicking on the X in the upper-right corner of the window.

9. From the desktop, right-click on the Shared Docs folder and select Sharing.

10. From the Shared Properties window, select the Shared As radio button.

11. Note that Shared Docs now appears in the Share Name field.

12. On the same page, change the Access Type from Read Only to Full; then click on OK to exit.

When you return to the desktop, you will notice that a hand appears under the Shared Docs folder. This indicates that the resource is available on the network.

Part 6: Testing the Configuration

1. On the PC without the shared directory, double-click on the Network Neighborhood icon on the desktop.

   You should see the PC with the Shared Docs folder listed in the Network Neighborhood window.

2. Double-click on PC with Shared Docs to see the Shared Docs folder.

3. Double-click on the Shared Docs folder to see the shared_document.txt file.

4. Double-click on to open the shared_document.txt file; then close it without making changes.

5. You can add a new document to the Shared Docs folder by right-clicking and selecting Add, New Text Document. Follow Part 5 to create a document.

6. Return to the PC with the Shared Docs folder and open it to reveal your new document.

Congratulations, you did it!

Conclusions

All computers that need to communicate for the purpose of sharing resources and data must do so using the same protocols. NetBEUI is one such protocol that enables computer users to easily share data. Because of its widely accepted status and ease of setup and implementation, NetBEUI is a popular choice for home networking.

Telecom Demarc

INTRODUCTION

At one time, the telephone company took complete responsibility for the telephone, its installation, and its performance. As the demand for consumer choice increased and as residential installations became more complicated, the telephone company scaled back its level of support to a point near where the wire from the central office first touches the residence. This is called the "point of demarcation." It is the responsibility of the installer to know when the telephone company network is at fault and when there is a problem within the customer premises. The device terminating the wire at the demarcation point is called the Network Interface Device, or NID. In addition to providing access to the wire, the NID also provides lightning protection for the wire entering the building.

WHAT YOU WILL DO

In this lab, you use a telephone to determine whether a problem exists in the telephone network or in the residence.

WHAT YOU WILL NEED

▶ On the HTI lab wall, obtain access to the ADO point on the telephony distribution frame.

▶ Obtain the Paladin LAN Pro Navigator test set and amplified probe.

▶ Obtain one three-to-six foot piece of Category 5e cable, terminated on one end with an 8P8C (RJ-45) connector.

▶ Obtain one short piece (10 inches to 20 inches) of Category 5e cable, terminated on one end with an 8P8C (RJ-45) connector. Fan the ends out in a North-South-East-West pattern, so that each pair is easily accessible.

Setting Up

For this lab, we will assume that the telephone distribution module within the lab wall is the telephone company point of demarcation. There are many similarities, but you may want to repeat this exercise at an actual point of demarcation when you have one available.

Part 1: Connecting to the Demarc

1. Terminate one end of a one- to two-meter piece of Category 5e cable with an 8P8C (RJ-45) plug.

2. Simulate the telephone company service drop by running the other end of this cable to the incoming network punchdown block, which is near the top left of the telephony distribution module (see Figure 4-1-1).

Figure 4-1-1 Run the cable to the incoming network punchdown block.

3. Punch down the wires in their proper order. The punchdown sequence you should use, left to right, is the following:

 a. White-Blue/Blue

 b. White-Orange/Orange

 c. White-Green/Green

 d. White-Brown/Brown

4. Plug the 8P8C connector created in step 1 into the LAN Pro Navigator.

5. Turn on the LAN Pro Navigator by pressing the On button once.

6. Set the device into tone generator mode by pressing the Tone button and holding it for about five seconds until the tone light comes on and the light labeled 1/2 begins blinking.

Part 2: Testing the Distribution Board

1. The tone generator is now sending tone down the cable only on line 3 (pins 1 and 2). Press the Tone button twice more so that the 4/5 light is blinking. Pins 4 and 5 are the pins for telephone line 1 (the blue pair). Configuring the tone generator to send tone into the incoming network punchdown block on pins 4 and 5 simulates an active telephone with a single line.

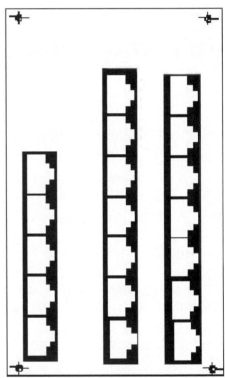

Figure 4-1-2 Check the jacks on the distribution board.

2. Use the amplified probe to check each one of the jacks on the distribution board. Indicate which jacks are "live," that is, receiving tone, by writing the number "1" in each jack of Figure 4-1-2.

3. Press the Tone button on the tone generator four more times. This cycles the tone to pins 3/6, which is line 2 (the orange pair).

4. Use the amplified probe to check each of the jacks on the distribution board. Indicate which jacks are "live," that is, receiving tone, by writing the number "2" in each jack of Figure 4-1-3.

5. Press the Tone button on the tone generator four more times. This will cycle the tone to pins 1/2, which is line 3 (the green pair).

6. Use the amplified probe to check each of the jacks on the distribution board. Indicate which jacks are "live," that is, receiving tone, by writing the number "3" in each jack of Figure 4-1-4.

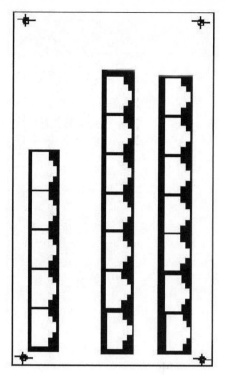

Figure 4-1-3 Check the jacks on the distribution board.

7. Press the Tone button on the tone generator three more times. This cycles the tone to pins 7/8, which is line 4 (the brown pair).

8. Use the amplified probe to check each of the jacks on the distribution board. Indicate which jacks are "live," that is, receiving tone, by writing the number "4" in each jack of Figure 4-1-5.

9. Using your knowledge of the distribution board, indicate with the numbers 1–4 which telephone line will appear at each jack on the board in Figure 4-1-6 (the first answer has been provided).

The jacks can now be used to distribute different telephone lines to different locations throughout the home.

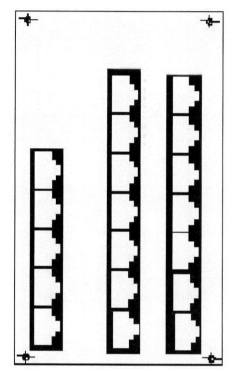

Figure 4-1-4 Check the jacks on the distribution board.

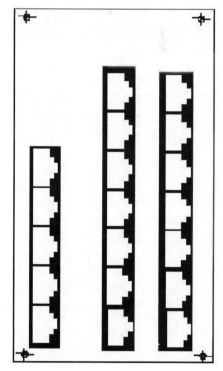

Figure 4-1-5 Check the jacks on the distribution board.

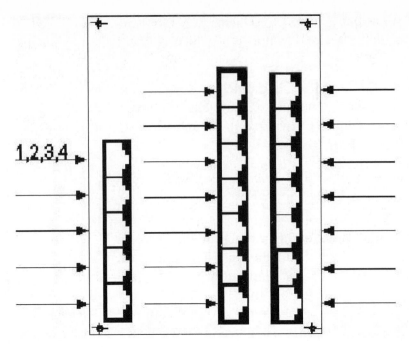

Figure 4-1-6 Indicate which telephone line appears at each jack on the board.

Part 3: Testing the ADO

1. Examine the jack field located on the telephony distribution module. It contains three columns of jacks, with the one in the center and the one on the right being full height (eight jacks). The left is interrupted by the demarc input and the Security jack.

   At the top of the center and the right are two jacks, marked Test and ADO, respectively. There should be a short jumper wire that bridges these two jacks.

   The wires that come in on the punchdown block marked FROM DEMARCATION pass directly to the TEST jack on the board. The purpose of this jack is to allow testing of the incoming line (see Figure 4-1-7).

   The TEST jack jumps to the auxiliary disconnect outlet (ADO) through the short jumper cable (see Figure 4-1-8).

   From the ADO jack, the signals are sent to the rest of the jacks on the board.

2. Set the control unit of the LAN Pro Navigator to produce tone on all pairs at once by clicking the tone button until all the pair lights light up.

3. Using the amplified probe, probe around the demarc cable punchblock. Do signals appear to be present on all wires? _____

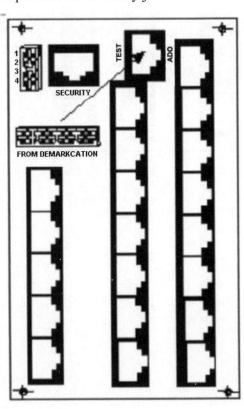

Figure 4-1-7 TEST jack

4. Remove the plug from the ADO socket. Insert the probe of the cable tester into the jack and move it around a little. Do all the signals appear to be present? _____

5. The ADO has a direct connection to the demarc punchdown. To verify this, restore the jumper between the ADO and TEST jacks.

6. Insert the probe into any of the jacks in the left row of the distribution board. How does the behavior of the jacks in this row compare to the ADO jack, keeping in mind that there may be minor variations in each pair and each jack?

_____

_____

7. Remove the jumper between the ADO and TEST jacks (lifting one end out of its plug should be sufficient). Repeat the measurements in the left column of jacks. What is different?

_____

_____

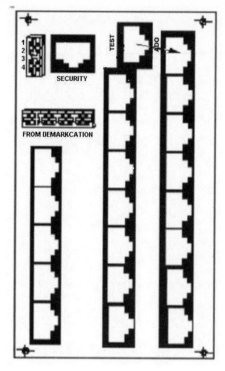

Figure 4-1-8 TEST jack jumps to the ADO jack.

Conclusions

The use of an on-card demarc increases the flexibility of the Structured Media Center by allowing you to pull the Line Distribution Module out of circuit. With the demarc isolated, the incoming line can be tested for signal. If the line is dead, even with the ADO test point opened, the problem is likely to be within the telephone company's network.

Line Distribution

INTRODUCTION

Passing all of the lines from the demarc, through the ADO, and out the jacks on the left column is a straightforward form of line distribution that would work in many home offices with multiline phones. In many cases, however, the phones to be connected may be single-line or two-line phones. These phones are designed to obtain their connection on pair 1 (pins 4 and 5). A two-line phone will be designed to have its primary line on pair 1 (pins 4 and 5) and its secondary line on pair 2 (pins 3 and 6).

What happens when you want to connect a single-line phone to the third line coming in from the demarc? There needs to be some way to connect line 3 (pins 1 and 2) from the demarc to line 1 (pins 4 and 5) on the cable that goes to the phone. This is the advantage of the Line Distribution Board.

WHAT YOU WILL DO

In this lab, you patch cross-connects between the ADO point on the Line Distribution Board and wires that lead to work area outlets from the Structured Media Center (SMC).

WHAT YOU WILL NEED

▶   Access to a Structured Media Center

▶   Line Distribution Board and Category 5 or Category 5e distribution board

▶   Paladin LAN Pro Navigator and Cable Probe

Setting Up

Configure the Line Distribution Board with the Paladin LAN Pro Navigator feeding the demarcation point (this should already have been connected in the previous lab). Have several patch cables available for cross-connection.

Part 1: Basic Line Distribution

The Line Distribution Board makes line signals available in various combinations. Eight of the jacks on the Line Distribution Board are configured to deliver the input lines to the demarc to the rest of the distribution system, pair-for-pair. These are located in the left column and at the top of the center column.

Under the TIA/EIA 570-A standard, each room that is wired must be served by at least one work area outlet that has at least one unshielded twisted-pair cable and one coaxial cable. These work area adapters are to terminate in the SMC. Looping from connector to connector (so-called "daisy chaining") is not allowed under the standard. For this reason, the wires used in the installation all terminate in a series of distribution boards. These boards convert the incoming four-pair cable into an 8P8C connector for getting the signal from the main Line Distribution Board. Each jack corresponds to one of the 110 connectors on the board. For example, Jack number 1 corresponds to the 110 connector on the top right of the board (see Figure 4-2-1).

Examine the distribution board and take careful notice of each jack. Note the different numbering patterns (see Figure 4-2-2).

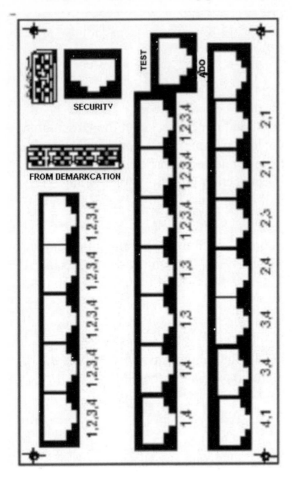

Figure 4-2-2 Note the different numbering patterns.

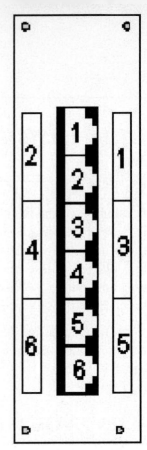

Figure 4-2-1 Each jack corresponds to one of the 110 connectors on the board.

Part 2:
Cross-Connecting

When accommodating a simple single- or dual-line telephone, the desired line from the demarc must enter the phone on its line 1 (or lines 1 and 2 in the case of dual-line telephones). This can be accomplished by connecting a jumper between the CAT 5e module and the Line Distribution Board at one of the seven jacks that provide only two lines. The numbers used—such as "4,1"—indicate that the fourth pair on the demarc will show up as the first pair in the jack, and the first pair in the demarc will show up as the second pair in the jack.

1. Demonstrate distribution by placing a jumper from the 4,1 jack into an available jack on the line module (see Figure 4-2-3).
2. Set the tone generator to put a tone on the fourth pair (white-brown/brown) of the demarc.
3. Using the tone probe unit, trace the positions on the corresponding punchdown block (this is the 110 block on the lower left of the board, #6).
4. When you move the probe near the output punch block corresponding to the line module jack, which pair do you find has the strongest sounding signal?_____
5. Now, change the tone generator to put tone on the first pair (white-blue/blue). Listen again with the probe near the punch block. Which pair seems to be the strongest?

_____

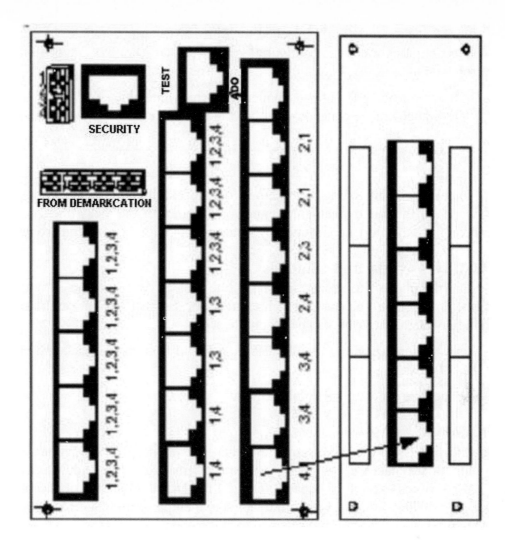

Figure 4-2-3 Place a jumper from the 4,1 jack into an available jack on the line module.

The process of moving a signal from one pair on one side of a punch block or patch panel to a different pair on the other side of the punch block or patch panel is known as cross-connecting.

Part 3: Advanced Line Distribution

Wire Pair Color	Line #	Pair #	Pins
White-Blue/Blue	Line 1	Pair 1	Pins 4 and 5
White-Orange/Orange	Line 2	Pair 2	Pins 3 and 6
White-Green/Green	Line 3	Pair 3	Pins 1 and 2
White-Brown/Brown	Line 4	Pair 4	Pins 7 and 8

1. Count the number of jacks on the Line Distribution Board that serve line 2 on the first pair and line 3 on the second pair. How many are there? _____

    Where is this jack located? _____ column; _____ from the top.

2. The 2,3 jack can easily be patched to a line distribution module, but what if a second phone requires a similar configuration? To create a second 2,3 configuration, jumper from one of the 1,2,3,4 jacks to the desired line module.

3. On the line feeding the phone, connect the white-blue/blue (first pair to the phone) to the white-orange/orange punchdown point. Line 2 now feeds line 1 on the phone. Why?

_____

_____

4. Connect the white-orange/orange pair on the cable going to the phone (line 2 to the phone) to the white-green/green punchdown point of the jack (line 3). Verify that you have made the proper connection by ringing it out with the tone and probe unit.

Part 4: Moves, Adds, and Changes

Imagine that the phone that needed incoming line 2 as its first line and incoming line 3 as its second line now needs to be changed to incoming line 1 as its first line, and incoming line 4 as its second line. Here's how this can be accomplished:

1. Move the jumper on the Line Distribution Board from the 2,3 jack and place it in one of the 1,4 jacks.

2. Use the tone generator and probe to verify.

Conclusions

Cross-connecting is used to serve single- and dual-line phones from multiline sources. The Line Distribution Board makes this simple by almost automating moves, adds, and changes. When requests for services exceed the capacity of the Line Distribution Board, acceptable substitutes can often be provided by using an all-line jack and then performing a cross-connect at the point where the wire leaves the SMC and heads toward the phone.

Using an RJ31X

INTRODUCTION

The RJ31X jack, which is used to intercept a telephone line for a security system, is used as an intermediary connection point between the telephone service provider and a user outlet. When there is no security system plugged into the RJ31X, internal shorting bars allow dial tone to pass directly to the user outlet. Inserting a plug into the jack raises the shorting bar and prevents dial tone from going directly to the user outlet. The security system accepts dial tone from pins 4 and 5, and sends dial tone out to the RJ31X on pins 1 and 8. The user outlet functions in the normal manner. The security system can now interrupt the user outlet and seize the line in the case of a security event. Figure 4-3-1 shows the wiring schematic for the RJ31X.

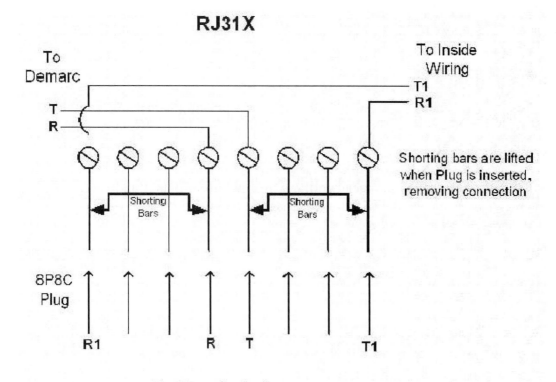

Figure 4-3-1 RJ31X wiring schematic

WHAT YOU WILL DO

In this lab, you will examine the screw terminals of an RJ31X with a multimeter to discover how it works.

WHAT YOU WILL NEED

‣ RJ31X surface mount jack

‣ Corded RJ-45 plug

‣ Multimeter

Setting Up

1. Remove the cover from the RJ31X.
2. Set the multimeter to ohms.

Activity

1. Examine the RJ31X. Look carefully at the rear of the jack. Notice how the pins contact the shorting bar. Trace all of the wires from the screw terminals to the jack pins.
2. While looking carefully at the rear of the jack, insert the plug. What happens to the pins when the plug is inserted?

   _____

3. Remove the RJ-45 plug. What happens to the pins when the plug is removed?

   _____

4. Follow all of the wires to their pin locations. Connect the leads of the ohmmeter to the wires that go to pins 5 and 8. The wire for pin 5 should be red, whereas the wire for pin 8 should be gray (this may vary with different jack styles). What is the reading on the meter? _____

5. Connect the leads of the meter to the wires that go to pins 4 and 1. The wire for pin 4 should be green and the wire for pin 1 should be brown (this may vary with jack styles). What is the reading on the meter? _____

6. Insert the plug into the jack. Measure the wires that go to pins 5 and 8. Measure the wires that lead to pins 4 and 1. What are the readings on the meter?

   _____

Conclusions

Dial tone from the telephone service provider comes into the RJ31X on the red/green pair terminated on the screw terminals for pins 4 and 5. Dial tone leaves the RJ31X jack on the screw terminals for pins 1 and 8.

When no plug is installed, dial tone passes directly from pins 4 and 5 to pins 1 and 2. When a plug is installed, dial tone passes from pins 4 and 5, passes through the security system, and goes back to pins 1 and 8.

This allows the security system to interrupt the line if it is in use.

Exploring Coaxial Cable

INTRODUCTION

Coaxial cable is often used in high-bandwidth applications and in applications in which the signal needs to be shielded from outside interference or in which the signal causes outside interference. The shielding keeps interference (noise) out of the cable and the signal inside the cable.

Most circuits require two conductors to function. In a coaxial cable, the center conductor traditionally carries the signal, whereas the shield acts as the ground path. Although this ground path may not always be required for signal transmission, it may be required for signal shielding.

A coaxial cable is a high-performance cable. The center conductor must be accurately centered in the cable for maximum performance. Connectors for coaxial cable keep the center conductor centered and isolated from the shield.

Different shielding types and shielding density affect the bandwidth that the cable can carry. The coaxial cable normally found in a home has a copper or copper-clad steel center conductor and a number of layers of braid. Basic Series 6 cable has a layer of foil covering and insulating material called the dielectric. A braid is constructed over the layer of foil. Quad-shield cable has the same layer of foil over the dielectric, a layer of braid over the foil, a second layer of foil over this braid, and a second braid over the second layer of foil.

High-bandwidth signals tend to travel over the outer edges of a wire. This is known as *skin effect*. Because the signals travel over the outer circumference of the wire, it is acceptable to construct the wire of steel and coat the steel strand with copper. The steel lowers the cost and provides a stronger cable. The copper coating provides a low-resistance path for the signal.

WHAT YOU WILL DO

In this lab, you strip the layers from a Series 6 coaxial cable and examine the construction of the cable.

WHAT YOU WILL NEED

▶ Section of Series 6 quad-shield coaxial cable

▶ Cable-cutter and coaxial cable stripper

Activity

1. Place the coaxial cable stripper approximately 1½ inches from the end of the cable (see Figure 5-1-1).

Figure 5-1-1 Placing the Coaxial Cable Stripper

2. Rotate the stripper (in the direction indicated on the stripper) around the cable two or three times (this is called *ringing the cable*).

    The stripper has two blades: One blade cuts through the outer jacket, braiding, and dielectric—allowing you to expose the center conductor; the other blade cuts through the outer jacket only.

3. After ringing the cable, pull off the outer jacket and the section that contains the center conductor dielectric (see Figure 5-1-2).

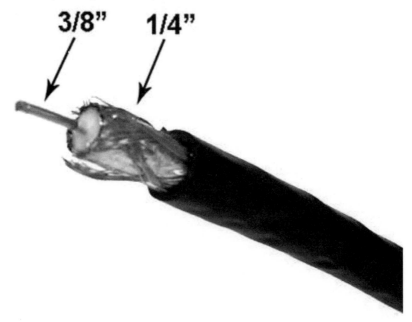

Figure 5-1-2 Stripped cable

4. Peel back the different layers of braid and foil. Notice the fine wires that make up the braid. What would happen if even one of these wires came into contact with the center conductor?

    _____

    _____

5. Bend the center conductor with your fingers. Notice how stiff the center conductor is compared to a common copper wire of the same gauge. What advantages does a steel center conductor provide?

    _____

    _____

6. Why is this cable called quad-shield?

    _____

    _____

Conclusions

Higher-performance coaxial cables have denser braids and more of them. Older coaxial cables used in home video distribution were an RG59 single-braided cable. Higher-performance cables in integrated homes are recommended to be Series 6 cable with a quad-shield.

RG-6 Connectors

INTRODUCTION

The cable used in Leviton structured cabling systems is Series 6 quad-shield (formerly called RG-6 Quad). Traditionally, an RF video signal uses a type F connector. There are many types of F connectors. The older style F connector (still commonly used) is crimped on the outer sleeve of the connector, which leaves a six-sided crimp (this is called a hex crimp connector).

The newer style of connector, often referred to as a *snap-n-seal connector*, has a ring that is pushed up into the connector by a special crimping tool. This connector provides perfectly concentric pressure on the cable, and also has the advantage of being watertight. The snap-n-seal connector also features an O-ring between the nut of the connector and the barrel, which keeps water out of the connection.

WHAT YOU WILL DO

In this lab, you learn how to install a snap-n-seal F connector and test the terminated cable.

WHAT YOU WILL NEED

▸   Six-foot length of Series 6 quad-shield cable

▸   Snap-n-seal connector

▸   Coax stripping tool

▸   Paladin coax crimping tool

▸   Lan Pro Navigator with BNC-to-F adapters.

Part 1: Stripping the Cable

1. Use the coax stripping tool to remove the jacket and appropriate layers at the end of the cable. The tool has two blades: One blade has a small notch; the other blade is straight and slightly recessed. The blades are spaced to remove the proper amount of outer jacket and dielectric.

2. Open the tool and slip the cable into the tool. The end of the cable should be toward the blade with the notch, and there should be a small arrow on the tool to show the direction of the cable. Place the tool approximately one inch from the end of the cable.

3. Rotate the tool a few times (there is an arrow on the tool to show the direction of rotation) until the blades no longer cut material.

4. Remove the tool and pull off the braid, dielectric, and outer jacket.

Part 2: Installing the Connector

1. Push a snap-n-seal connector onto the end of the cable. A slight twisting action to seat the connector may be necessary.

2. Look into the end of the connector. Make sure the dielectric is flush with the inner body of the connector (see Figure 5-2-1).

Figure 5-2-1 The dielectic flush with the inner body of the connector

Part 3: Crimping the Connector

1. Insert the assembly into the crimping tool (see Figure 5-2-2). The end of the center conductor and the end of the connector go into a cavity in the tool.

2. Snap the cable and back end of the connector into the retaining arms of the tool.

3. Crimp the connector. Watch how the tool pushes the tail of the connector together.

4. Remove the crimped connector and strip off any excess center conductor. The center conductor should protrude approximately one-quarter inch from the crimped-on connector.

Figure 5-2-2 Crimping tool

Part 4: Testing the Cable

When both ends of the cable are terminated, test the cable:

1. Connect one end to the Lan Pro Navigator main unit, the other end to the remote unit.

2. Turn the tester on and test the cable.

If the cable does *not* pass the test, one of the connectors has been installed incorrectly. Do the following:

1. Examine each connector and the test results. The tester should show some type of cable fault on one end.

2. Cut off this connector, restrip the cable, reinstall a connector, and retest.

Conclusions

The snap-n-seal connector is a fast, efficient way to terminate coaxial cable. This type of connector provides a sturdy termination that is both solid and weatherproof. The tool to crimp the connector is very easy to use and helps to make the installation of the connector quick and simple.

Video Adapters

INTRODUCTION

There are two ways in which most modern TVs get their signals:

▶ The audio and video outputs (the ones that use RCA connectors) carry baseband audio and video signals. The video signal is usually one volt peak to peak. The audio signal is at "line level." These signals can be sourced from a camcorder, a VCR (using the video output, not the channel 3/4 output), a video game machine (using the audio/video output, not the F-connector output), and some computer cards.

▶ The modulated RF connector is used for a TV signal as opposed to a video signal. This signal can be sourced by an off-air antenna, the output of a basic cable system, and the channel 3/4 output of a cable TV set top box.

Because these signals differ, so do the connectors they use. Modulated RF nearly always uses a type F connector. Audio and video frequently use RCA plugs. (Professional video and test equipment frequently use BNC connectors.) It is a tenet of structured cabling systems that the number of types of media should be standardized and routed through a common wire-management system. For this reason, this course is standardized on RG-6, quad-shield, for video cabling. This 75-ohm cable is suitable for a single channel of baseband video or a broadband cable TV system signal.

To use the same cable in all these applications requires adapting the type F fitting, which is required for RF to accommodate the various other connectors available.

WHAT YOU WILL DO

In this lab, you learn the differences between video and modulated RF and connect a video signal two different ways.

WHAT YOU WILL NEED

▶ Cable-ready TV

▶ Video signal source, such as a VCR or a DVD player

▶ Six-foot length of Series 6 quad-shield cable with F connectors on both ends

▶ Six-foot length of Series 6 quad-shield cable with F-to-RCA adapters on both ends

Safety

The voltages supplied by the TV and VCR signal input and output connectors should be so low as to not likely be dangerous, *assuming that the equipment is operating correctly*. If a ground fault or ground leakage is present, the situation changes. An appliance with a known ground problem should be referred to a qualified service technician. Errant line voltages can be deadly.

Part 1: Preparing the TV

1. Adjust the TV to a channel over which it can view the VCR—most likely, channel 3 or 4.

2. Also, locate the video inputs on the TV, as shown in Figure 5-3-1 (they probably use RCA connectors).

Part 2: Preparing the Video Source

The video source in your lab may be a VCR or a DVD player. These devices have a common goal: They produce a signal that can be displayed on a standard television set.

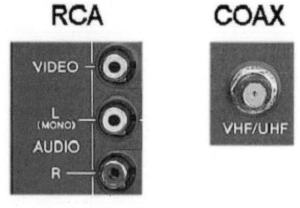

Figure 5-3-1 Video inputs on the TV

1. Determine if channel 3 is operating in your area.

2. If it is, set the switch to channel 4; if it is not, set the switch to channel 3.

   To which channel should the switch be set for a classroom in your area? _____

Part 3: Identifying Adapters

Locate several of the adapters that enable the type F connectors on the quad-shield cable to mate with RCA and BNC connectors so that the cable can be used for additional purposes.

What kind of appliance can you use to demonstrate the use of a type F-to-BNC adapter?

_____

What kind of appliance can you use to demonstrate the use of a type F-to-RCA adapter?

_____

Part 4: Connecting a Video Signal with RF

1. On the video source, connect a short piece (approximately six feet) of coaxial cable between the video source RF output and the TV RF input (using F connectors).

2. Activate the video source by turning it on or starting a tape rolling. Tune in the signal on the TV.

3. Take careful note of the picture:

   ▶ Are the colors accurate? _____

   ▶ Are the colors lifelike? _____

   ▶ How is the detail? Notice any particular areas that are rich in vertical and horizontal lines and any letters or shapes, and how well they appear onscreen. Describe these areas of the image. _____

   _____

   ▶ Note the audio quality. Is it clear? _____ Is there any hiss or buzz? _____

Part 5: Connecting a Video Signal with Baseband

1. Connect the video source to the TV using the video and audio cables. (For the purpose of this step, use a coaxial cable with F-to-RCA adapters fitted to the ends of the cables, as shown in Figure 5-3-2.) Make sure to connect the cable between the video output of the source and the video input of the destination.

2. Take careful note of the television picture:

   ▶ Are the colors accurate? _____

   ▶ Are the colors lifelike? _____

Figure 5-3-2 F-to-RCA Adapter

   ▶ How is the detail? In particular, notice areas that are rich in vertical and horizontal lines and any letters or shapes, and how well they appear onscreen. Describe these areas of the image.

_____

Part 6: RF versus Baseband

Generally speaking, which of the two images (the one using RF or the one using baseband video) is more crisp and clear? _____

What do you think is the reason for this situation?

_____

_____

Is it possible to connect a baseband signal source to an RF video input?

_____

What type of connector is normally used by a baseband signal? _____

What type of connector is normally used for an RF video signal?_____

1. Select the proper cable and make a baseband connection from your video source to the TV.

2. Select the proper cable and make an RF video connection to the TV.

Conclusions

There are many differences between transmitting baseband and RF video. Even though these transmission types are different, the connectors can be interchanged according to the transmission type. The connectors adapt the media to accept the transmitted signals. Both of these transmission types can use coaxial cable. For better quality video, a dedicated baseband signal is preferable.

RF Distribution

INTRODUCTION

This lab explores the techniques and principles involved with converting composite video to a modulated RF signal.

The cables and configurations from the previous labs in Chapter 5 are necessary to properly complete this lab. If you have not done the previous Chapter 5 labs, please do them before continuing with this one.

The Structured Media Center Modulator takes the signal from a video device such as a security camera and modulates the signal to one unused cable TV channel, which can then be combined with the CATV signal to multiple locations anywhere in the home. This effectively provides a whole-house security channel.

The Structured Media System Modulator, when used with Structured Media Video Distribution modules such as a 3x8 bi-directional video module, must be used with a Notch Filter to clear the target cable channel and avoid interference in a Structured Media System.

WHAT YOU WILL DO

In this lab, you configure the video modulator to convert composite video to RF and determine what frequency range the notch filter blocks.

WHAT YOU WILL NEED

- ▶ Television
- ▶ Output from a video camera
- ▶ Video modulator
- ▶ Notch filter
- ▶ 1 RG-6 coax patch cable with F connectors on each end

Setting Up

1. Identify the video modulator and verify that it has power.
2. Position the television near the SMC where the video modulator is mounted.
3. Identify the coax cables coming from the video camera.
4. Locate the notch filter.

Part 1: Configuring the Video Modulator

A *video modulator* is a device used to convert composite video to a modulated RF signal capable of being received through a television or VCR tuner (see Figure 5-4-1). The modulated video signal can also have an audio signal added to it.

Figure 5-4-1 Video modulator

Is there a familiar household electronic device that also has a built-in video modulator?

A VCR has video-in and audio-in and has modulated RF output. Most VCRs allow the user to select the output channel (either 3 or 4) through a selector switch. The modulator we will be using also has switches to set the output channel.

1. Remove the access plate from the bottom of the video modulator, as shown in Figure 5-4-2 (unscrew one of the screws to remove the access plate).

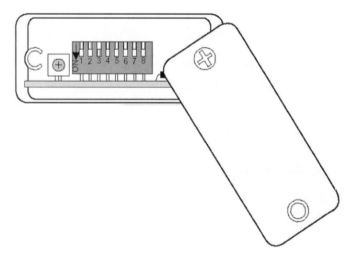

Figure 5-4-2 Removing the access plate

2. Locate the eight-position DIP switch inside and notice the numbers on the circuit board under each individual switch (see Figure 5-4-3).

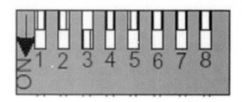

3. A switch in an on position adds that value to the channel number; an off switch does nothing. The total of the values for all the on switches equals the output channel number of the modulator (see Figure 5-4-4).

Figure 5-4-3 Each switch is numbered.

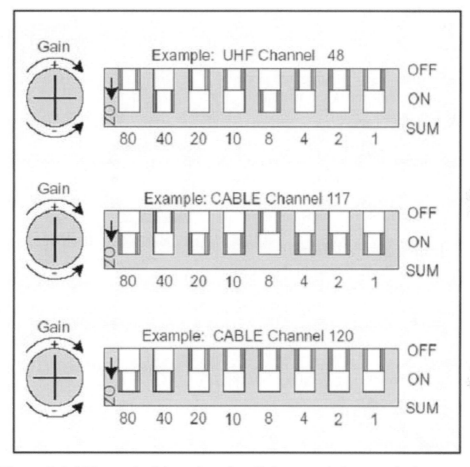

Figure 5-4-4 The total of the values for all the on switches equals the output channel number of the modulator.

If the switches corresponding to 40, 20, 10, and 2 are all set to on and all others are set to off, what channel will the output be?

_____

4. Set the switches with the preceding values. Do not replace the cover at this time.

Part 2: Connecting the Video Modulator

1. Connect the coax with an F-to-RCA adapter from one of the cameras to the video-in on the television.

2. Turn on the television and select video-in to verify that the camera signal is present on the coax.

3. Move the camera's coax from the back of the television to the video-in on the modulator. (Remember, composite video uses RCA connectors, and modulated RF uses F connectors.)

4. Connect a coax patch cable from the video modulator output (F connector) to the antenna-in on the television (see Figure 5-4-5).

5. Verify that the modulator is connected to its power supply.

6. Change the television to select the tuner and set the channel to the one configured in step 1. The picture from the camera should be displayed on the television.

If there is no picture, it may be necessary to start over from the beginning of Part 1. If you can locate the problem, ask the instructor for assistance. Do not proceed with the rest of the lab until you have successfully completed these two steps.

Part 3: Installing a Notch Filter

A *notch filter* is a device that blocks certain frequencies or ranges of frequencies while allowing others to pass through (see Figure 5-4-6). The television broadcast range uses frequencies from approximately 50 MHz to 900 MHz, with each channel occupying a 6 MHz "window." A notch filter can be used to block a single channel or range of channels.

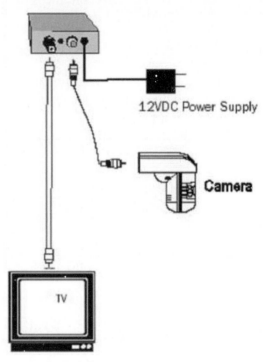

Video Modulator

12VDC Power Supply

Camera

TV

Figure 5-4-5 Attaching the cable to the television antenna

Male End

Female End

Notch Filter

Figure 5-4-6 Notch filter

Can you think of applications in which notch filters might be used?

_____

_____

_____

1. Examine the notch filter. Are there any indicators to show which end is the input and which is the output?

_____

_____

2. With the television displaying the picture from the camera, disconnect the patch cable from the output side of the video modulator (F connector).

3. Thread the male end of the notch filter onto the modulator's output connector (see Figure 5-4-7).

4. Attach the loose end of the coax patch cable that goes to the television to the female end of the notch filter.

5. Verify that the television is still displaying the picture from the camera.

Part 4: Determining Notch Filter Frequencies

Notch filters may block a single channel or range of channels, depending on the application. If the range of the filter is not known, you can determine it by using the variable channels in the video modulator.

1. Change the settings on the modulator DIP switches to the next higher channel.

2. Set the television tuner to the corresponding channel.

3. Advance the channels on the television together with the channels on the modulator one at a time until the picture no longer shows on the television.

   What channel did the notch filter first block out? _____

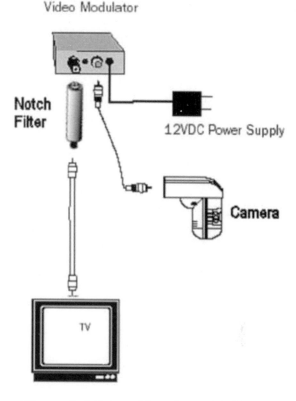

Figure 5-4-7 Attaching the notch filter onto the modulator's output connector

4. Continue advancing the channels on the modulator and television until the camera's picture returns to the television.

   What channel did the notch filter last block out? _____

Conclusions

Video modulators convert composite video signals into a modulated RF signal capable of being received by a standard television or VCR tuner. Modulators can also combine an audio signal with the video to form one signal.

Notch filters prevent certain frequencies from passing through. In the lab, we used a notch filter to block the frequency from one camera, simulating how a notch filter is used in a real installation. It can be installed in an incoming cable TV coax to block a range of channels that can be used by security cameras. The homeowner can then view the camera pictures from any television just by selecting the proper channel.

Video Splitters and Combiners

INTRODUCTION

Video splitters are used to distribute modulated RF signals to many locations throughout a home. In the old scheme of cabling, many small splitters were used to create a daisy chain wiring system. New standards define a star wiring scheme, in which all cables come from a central location. The daisy chain wiring used two-way splitters in many locations; the star distribution method uses a single large splitter in a central location. This central splitter may be a six-way, eight-way, or even 12-way splitter. A splitter has one in leg and multiple out legs.

Combiners are used to add or inject signals into a coaxial cable system. Combiners are actually splitters used in reverse. Multiple signals are connected to what is normally the out leg of a splitter. The signals must be on different frequencies so there is no interference. Notch filters can be used to clear occupied frequencies.

WHAT YOU WILL DO

In this lab, you run a video signal through a splitter to a TV and use a combiner to inject a video signal into the existing signal.

WHAT YOU WILL NEED

▶ Video splitter

▶ Camera, modulator, and TV from previous labs

Part 1: Video Signal Splitter

A video splitter usually has one input and from two to eight outputs. The splitter is used when more than one device (television or VCR) needs to receive a video signal, which is what typically happens when the incoming cable TV signal is divided between several rooms in the house. Figure 5-5-1 shows a video signal splitter.

This splitting has some signal strength loss associated with it. Signal strength is measured in dB, much like the way audio signal strength is measured. During the splitting process, the signal loses 3.5 dB for each output from the splitter. For example, in a three-way splitter, with an incoming signal strength of 19 dB, each output will have a strength of 8.5 dB $(19 - (3 \times 3.5))$.

Follow these steps:

1. Make sure the television displays the picture from one of the cameras.

2. Disconnect the patch cable from the antenna connection on the television and connect it to the input side of the splitter.

Figure 5-5-1 Video signal splitter

Figure 5-5-2 Video signal combiner

3. Use another patch cable to connect one of the splitter outputs to the antenna connection on the television. The camera picture should be displayed on the television.

4. Disconnect the patch cable from the output side of the video splitter and connect it to the other splitter output port. The television picture should look the same.

Part 2: Video Signal Combiner

A *video combiner* is like a video splitter in reverse. In fact, the same device may be used for each job. When using a splitter as a combiner, the labels *input* and *output* are reversed; you actually connect incoming signals to the output terminals. Figure 5-5-2 shows a video signal combiner.

Follow these steps:

1. Configure a second video modulator with video output from the second camera, as in step 2. Set the output channel to 77.

2. Configure the first video modulator to output channel 77.

3. Verify that each modulator is on the correct channel by setting the television on channel 77 and testing each output to see whether the camera picture is displayed on the television. Remember that channel 77 is in the range that is filtered by the notch filter—make sure it is not installed.

4. Connect a coax patch cable from the input terminal on the splitter to the antenna connection on the television. The splitter will now be used as a combiner, so the terminals now have opposite functions from their markings.

5. Connect each of the modulator outputs to the output terminals on the splitter. When the first one is connected, the camera picture should appear. When the second one is connected, the picture should become distorted. Why is the picture now distorted?

_____

6. Install the notch filter on the output terminal of one of the video modulators. Reattach the coax to the notch filter. Why does the picture on the television appear good again?

_____

Conclusions

Video splitters and combiners are devices that do exactly what their names imply. A splitter allows one incoming signal to be divided among two or more receivers. There is a loss of 3.5 dB per output during the splitting process. In a four-way splitter, total signal loss would be 4×3.5 for a total signal drop of 14 dB.

Video combiners are splitters in reverse. They combine two or more inputs into one output. The same device can be used to do both functions.

Video Camera Installation

INTRODUCTION

In this lab, you will connect two different types of video cameras to the Structured Media Center (SMC). The first camera is designed for exterior use, has audio as well as video capability, and is easily identified as a camera. The second camera is designed for inside installations, has no audio, and looks like a motion-sensing light switch.

One thing to notice in this lab is that the camera output is composite video—not a modulated RF signal. Composite video signals cannot be viewed in the normal television channel range without conversion. Many televisions and VCRs have video inputs that allow the camera to be connected directly to them.

WHAT YOU WILL DO

▶ Connect a video cable from the camera to the television.

▶ Connect a CAT 5 cable for audio and power from the camera to the SMC.

▶ Connect power to the CAT 5 cable in the SMC.

▶ Test to see that the camera functions properly.

WHAT YOU WILL NEED

▶ Television with video-in RCA jack

▶ Cameras

▶ RG-6 video cable with adapters for F-to-RCA male plugs

▶ CAT 5 cable long enough to reach from the camera to the SMC

▶ CAT 5 six-port voice and data module in the SMC

▶ Adapter cable: 8P8C plug to white RCA male audio

▶ 12 VDC distribution board in the SMC

▶ Punchdown tool with 110 blade

▶ Label maker

OPTIONAL ITEMS

Multimeter for verifying voltage and continuity

Setting Up

1. Place the television near the SMC containing the video sequencer and video modulator.

Figure 5-6-1 Exterior camera (left) and Decora interior camera (right)

2. Locate the two cameras that you will connect: Exterior (brown-colored one) and Decora interior (white-colored one that looks like a motion-sensing light switch) (see Figure 5-6-1).

3. Verify that the SMC has a CAT 5 six-port module and 12 VDC distribution board.

Part 1: Installing RG-6 Coax Cable

1. Route the video cable from the camera to the SMC. This cable should be long enough to reach the SMC as well as the video input on the television.

2. Put F-to-RCA male adapters on each end of this cable.

3. Connect the camera video out (yellow RCA jack on a pigtail) to your cable.

4. Label the other end of the cable to identify which camera it comes from.

5. Connect it to the video-in on the television.

Part 2: Installing CAT 5 Cable

1. Route a piece of CAT 5 cable from the camera to the SMC (it needs to reach the CAT 5 six-port data module).

2. Strip about two inches of the outer sheathing from the camera end of the cable.

3. Locate the orange pair of wires (and orange on the interior camera) and wrap them back around the outside of the sheathing—they are not used for this application but can be saved as spares.

4. The remaining pairs of wires should have one-half inch of insulation removed from each wire, and then the ends of each pair should be twisted together to make one wire out of each pair. This effectively increases the wire gauge.

5. Connect the wires to the terminals for the camera. The wiring code is as follows:

   ▶ Brown: 12 VDC + (this is the RED or Power terminal)

   ▶ Green: Ground (this is the BLK or Ground terminal)

   ▶ Blue: Audio (on the exterior camera) (this is the WHT terminal that is not marked)

How can the audio function with only one wire?

_____

_____

The other half of the audio circuit is supplied by the green ground wire. It acts as ground for both the power supply and the audio circuit.

_____

_____

Why does the video have two conductors?

_____

_____

Figure 5-6-2 The connector attached to the cable

The composite video signal has a dedicated ground in the shielding of the RG-6 cable. This signal is very susceptible to interference, so using the same ground as the power supply and audio could interfere with the video picture.

Part 3: Powering Off

Disconnect the power pack supplying the 12-volt power to the power distribution module in the SMC. If connections are made with the power on, the camera could be damaged!

Has the power pack been disconnected? _____ Yes _____ No

Part 4: Connecting to Punchdowns

1. Strip about two inches of sheathing from the end of the CAT 5 cable and wrap the orange wires back around the jacket.

2. Use the 110 punchdown tool to attach the brown, green, and blue (if audio is used) wires on one of the terminating strips of the CAT 5 data module following the color code on the module.

Part 5: Connecting the Exterior Camera with Audio

1. Locate the interface cable that came with the camera. It has an 8P8C plug on one end, two pairs of stripped wires, and a white RCA male plug for audio. This cable supplies the camera with power and also carries the audio from the camera.

2. Connect the brown pair of wires to the +12-volt contact of a Phoenix connector (the outside contact).

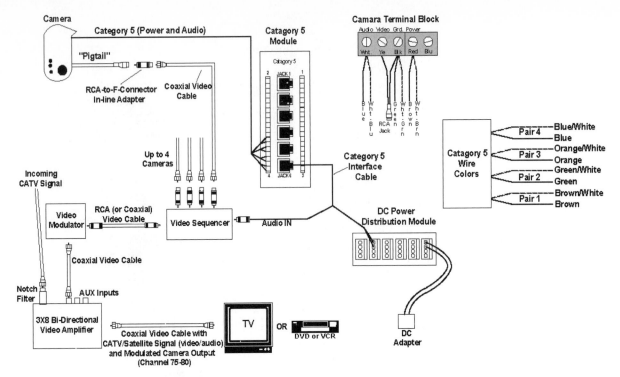

Figure 5-6-3 Outdoor camera wiring diagram

3. Connect the green pair to one of the ground terminals on the Phoenix connector (one of the inside contacts). Plug in the 8P8C.

4. Plug the white audio in connector from the interface cable into the left audio input jack on the television.

Part 6: Connecting the Decora Interior Camera, No Audio

1. Cut a piece of CAT 5 cable long enough to reach from the CAT 5 six-port voice and data module to the 12-volt power distribution board.

2. Strip about two inches of sheathing from each end of the cable and cut the blue and orange pairs off.

3. On the six-port end, use the 110 punchdown tool to attach the brown and green pairs of wire to one of the terminating strips.

4. Attach a label indicating that the corresponding 8P8C jack has 12 volts.

5. On the end near the 12-volt power distribution board, strip about one-half inch of insulation from the pairs and twist them together to form one brown wire and one green wire.

6. Connect the brown pair of wires to the +12-volt contact of a Phoenix connector (the outside contact).

7. Connect the green pair to one of the ground terminals on the Phoenix connector (one of the inside contacts).

8. Use a short 8P8C patch cable to jumper between the power supply and camera jacks.

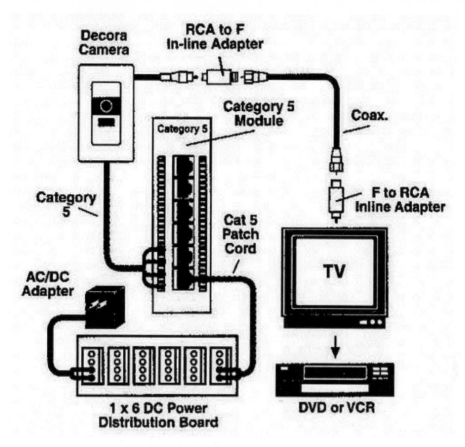

Figure 5-6-4 Interior camera wiring diagram

Part 7: Powering On and Testing

1. Plug the Phoenix connector onto the 12-volt power distribution board. DO NOT connect power until the instructor has verified that the wiring is correct!

   Did the instructor verify the wiring? _____ Yes _____ No

2. Reconnect the 12-volt power supply to the power distribution board. Turn on the television and set the input to be from the video source. You should see the picture from the camera. You can verify that each camera is working by unplugging the input from the back of the television and substituting the cable from the other camera.

Why is there sound for only one camera?

_____

The interior camera has no microphone to pick up sound. The exterior camera has a microphone and has been wired through to the SMC.

Conclusions

You have now successfully wired a camera to produce a video and audio signal capable of being viewed on a television or recorded on a VCR. The signal produced by the camera is a composite video signal—not one that be used through a television tuner. The video signals use coaxial cabling because they need more shielding from interference. The shield on the coax also prevents interference from originating in the video signal and traveling out to corrupt other adjacent signals.

Do not remove any of the connections you have made. They will be used in a future lab.

RF Amplifiers

INTRODUCTION

RF amplifiers can be used in the home in a number of different ways. The purpose of an amplifier is to strengthen or increase a weak signal. The inbound signal from the cable provider could be a weak signal. A good signal from the cable provider could be weakened by going through a large splitter or too many splitters. A single long run in a large home may have so much attenuation that the signal is too weak to be received when it gets to the television.

WHAT YOU WILL DO

▸ Simulate installing an amplifier where the cable signal is weak.

▸ Simulate installing an amplifier to feed a video signal to a cable run in excess of 150 feet.

WHAT YOU WILL NEED

▸ One video amplifier

▸ Four pieces of coaxial cable with F connectors installed, random lengths

▸ One six-way splitter

Setting Up

1. Install a video amplifier and a 1 x 6 video splitter into an SMC.

2. Label one of the coax cables "CATV service."

Part 1: Amplify a Weak Signal

1. Assuming that the CATV signal is weak, connect the cable labeled "CATV service" to the connector on the amplifier labeled "RF input" (Figure 5-7-1, video amplifier with one output). Cable TV or other video signals are run through this type of video amplifier before being split and distributed throughout the home.

Figure 5-7-1 Video amplifier

2. Connect a cable from the connector on the amplifier labeled "RF output" to the center connector on the six-way splitter. Figure 5-7-2 shows a six-way splitter

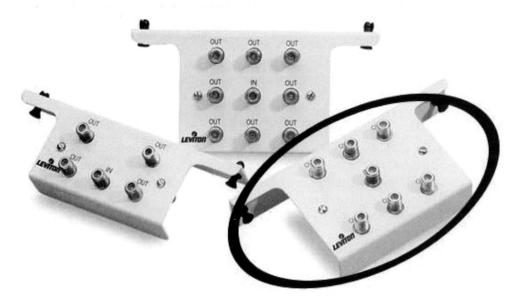

Figure 5-7-2 Six-way splitter

Part 2: Disconnect cables

Disconnect and remove all cables.

Part 3: Strengthen a Signal that Travels a Long Distance

1. Assuming that the CATV signal is adequate, connect the cable labeled "CATV service" to the center connector on the six-way splitter. Assume that one cable run will exceed 150 feet.

2. Connect a coax cable from one of the unused splitter ports to the RF input port on the video amplifier.

3. Connect a piece of cable representing the long cable run to the connector on the amplifier labeled "RF output."

Conclusions

Video amplifiers can be used to amplify the signal feeding all video outlets or a single outlet.

Amplified splitters can be used in place of single-line amplifiers. Modern CATV systems are two-way systems that support services such as digital cable, pay-per-view, and cable modems. All video amplifiers should be specified as two-way amplifiers.

DC Power Distribution

INTRODUCTION

Each Structured Media Center should be equipped with a power supply mounted on the bottom of the distribution panel. (This supply distributes 120 VAC and should be installed by a qualified electrician.) AC adapters, which plug into this power supply, provide low-voltage DC power to accessories within the SMC.

AC adapters can use up the limited outlet space in the SMC. A single AC adapter with a DC distribution panel can provide power to nearly all the active devices with the SMC. The function is similar to a patch panel. DC power can be simply patched to the devices requiring it.

WARNING: All electrical outlets and wires should be treated as if they are hot, whether they have been turned off or not.

WHAT YOU WILL DO

▸ Identify the power strip in the SMC and connect modular supplies to it, as required.

▸ Understand the use of the 48212-DCS 12 VDC power distribution module for situations requiring more complicated power distribution.

▸ Apply 12 V power to a section of CAT 5e cable.

▸ Test the cable for power and polarity.

WHAT YOU WILL NEED

▸ DC power distribution panel

▸ 1500 mA AC adapter

▸ Multimeter

▸ Small section of CAT 5e cable

Setting Up

Make sure that the outlets at the bottom of the SMC are powered by 120 VAC.

Part 1: Examine the DC Distribution Module

1. Examine the DC distribution module. Look at the bottom of the printed circuit board to see how all of the + voltages are in common with one another. The – voltages are also in common with one another. The module can also provide power through a coaxial cable.

Figure 5-8-1 DC Power Distribution Module

The small black connector at the left of the panel is the 12-volt DC input. This panel requires an AC adapter rated for 1500 mA.

The small green connectors supplied with the panel are called Phoenix connectors. They are connected to a wire and plugged into the distribution panel. Notice that the connectors can be plugged in only one way.

2. Plug a connector into a socket and remove it.

Part 2: Terminate CAT 5e Cable to the Connector

1. Mount the DC distribution panel in the SMC.

2. Take a short section of CAT 5e cable and remove about one inch of the jacket. Separate the four pairs.

3. Strip one-half inch of insulation from each conductor.

4. Take each pair and twist the ends of the bare copper wires together.

5. Do the same to the other end of the cable.

6. Look at the screws on the connector. Loosen each of the screws, but do not remove them.

7. Hold the connector with the connection to the distribution panel facing down.

8. Beginning at the left, insert the white/blue pair into the first hole under the screw.

9. Insert the white/orange pair under the next one, the white/green pair under the next, and the white brown/brown pair under the last.

10. Tighten all the screws.

Part 3: Make the Connections

1. Separate the pairs that are not terminated. Make sure that the bare wires do not touch one another or anything else.

2. Plug the Phoenix connector into the panel.

3. Plug the AC adapter into the 120 V outlet at the bottom of the SMC.

4. Plug the 12 VDC connector into the distribution panel.

   Did the small LED light?_____

Part 4: Measure Voltage

1. Obtain a multimeter and set it to "Volts." Figure 5-8-2 shows a Multimeter.

2. Carefully touch the red lead of the meter to the bare portion of the blue pair and touch the black lead of the meter to the orange pair.

   What is the reading on the meter? _____

   If the voltage were –12V, what would be wrong? _____

3. Unplug the voltage source from the distribution panel.

4. Cut the bare ends off of the CAT 5e cable.

5. Unplug the Phoenix connector from the distribution panel.

Conclusions

DC distribution panels provide a convenient method of powering devices in SMCs. The CAT 5e cable can be punched to a CAT 5e distribution panel and distributed to remote units throughout the house. Items such as video cameras would be powered in this manner.

The ends of the CAT 5e can also be twisted and terminated into another Phoenix connector to power other devices within the SMC. Devices such as the digital volume control interface are powered in this manner.

Figure 5-8-2 Multimeter

Decora Media System

INTRODUCTION

This lab demonstrates the capability to combine a composite video signal (security camera, VCR, DVD, satellite television) with stereo audio signals and transmit them up to 1000 feet through CAT 5 cable. This signal can also be repeated to as many as six receiving stations by utilizing an amplifying hub. If more than six receiving stations are needed, additional hubs can be daisy-chained off the primary hub.

WHAT YOU WILL DO

▶   Install Decora media system send and receive modules.

▶   Connect the units together with CAT 5 cable.

▶   Install a Decora media system hub.

▶   Connect the send and receive modules and redistribute the signal through the hub.

WHAT YOU WILL NEED

▶   Decora media system send and receive modules

▶   12 VDC power supply for send module

▶   Decora media system hub

▶   15 VDC power supply for hub

▶   Video and audio source (a VCR works well)

▶   Television with video-in and audio-in jacks

▶   Two CAT 5 patch cables terminated with 8P8C connectors

▶   Six RG-6 coax cables with RCA male plugs on them (or two molded video/audio/audio RCA [yellow—red—white] composite patch cables can be substituted)

OPTIONAL ITEMS

▶   Second receive module and television

▶   Three additional RG-6 coax cables and one additional 8P8C patch cable

Setting Up

1.   Position the VCR near the send module.

2.   Position the television near the receive module.

3.   Locate the media hub in the SMC and verify that it is getting power. Figure 5-9-1 shows a media hub.

4.   If a second television is used, locate it near the second receive module.

Refer to Figure 5-9-1 as you complete this lab.

Figure 5-9-1 Media hub

Part 1: Connecting the Send Module

1. Connect three patch cables from the video-out and audio-out jacks on the VCR to the video-in and audio-in jacks on the send module.

2. Connect a CAT 5 straight-through cable from the 8P8C connector on the send module to the 8P8C connector on the receive module.

3. Connect three patch cables from the video-out and audio-out jacks on the receive module to the video-in and audio-in jacks on the television.

4. Set the television to use video-in as the source of its incoming signal.

5. Plug in the 12 VDC supply to the send module.

6. Start the tape playing in the VCR.

The television should display the image and audio from the tape playing in the VCR. If the image is not showing, do the following:

1. Verify that the television is set to receive signals from its video-in port.

2. Verify that the VCR is playing a tape.

3. Verify that a straight-through CAT 5 cable was used.

4. Verify that the send module is getting 12 VDC power.

5. Verify that the audio and video connections are correct at each end.

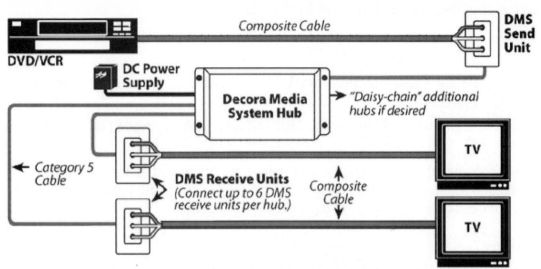

Figure 5-9-2 Schematic for setting up a Decora media system

Part 2: Distributing Signals through the Hub

1. Disconnect power from the Decora media system hub.

2. Disconnect power from the send module.

   Is the power disconnected? _____ Yes _____ No

3. Disconnect the CAT 5 cable from the send module and connect it to one of the six 8P8C output connections on the Decora media system hub.

4. Connect another CAT 5 patch cable from the send module to the 8P8C connector labeled input on the media system hub.

5. Reconnect power to the hub.

The VCR picture should once again be displayed on the television.

If the send module does not have power connected, how is it functioning?

The send module is getting power through the CAT 5 cable from the media hub, just as the receive module gets its power through the CAT 5 cable from either the hub or the send module.

Part 3: Connecting a Second Receive Module and Television

1. If there is more than one receive module, connect another CAT 5 straight-through patch cable between the receive module and another output on the media system hub.

2. Connect another set of audio and video patch cables between this receive module and the second television.

This television should receive the same picture quality as the first television. The hub acts as a splitter and amplifier, ensuring that the picture quality remains the same on all televisions.

Part 4: Connecting Input Directly to the Hub

1. Disconnect the CAT 5 patch cable that goes between the media system input and the send module.

2. Disconnect the audio and video patch cables from the input connections on the send module.

3. Connect these cables to the video-in and audio-in RCA connectors on the media system hub.

The television(s) should now display the output from the VCR, just as they did when the signal was sent through the send module. If the source of the video and audio is close to the media hub, the connections can be made directly to the hub.

Conclusions

The Decora media system send and receive modules can take composite video and audio signals from a VCR, security camera, DVD, or satellite television receivers and redistribute them up to 1000 feet through standard CAT 5 cabling. If just one device will be receiving the signal, a single receive module is all that is needed. If more than one device will be using the signal, a hub may be added to support a total of six outputs. The hub may also act as a send module and accept the composite video and audio inputs from the source device without using a separate send module. If more than six receiving stations are needed, additional hubs can be daisy-chained from the primary hub.

Audio Distribution

INTRODUCTION

The Leviton digital volume control interface allows the configuration of an audio zone in the home to provide quality audio, convenient volume control locations, and multiple speaker and volume controls within the zone.

The homeowner can enjoy centrally located audio sources that can be distributed throughout the home using speaker distribution modules. These modules allow numerous speakers to be connected to a source such as an audio/video receiver, a television, or even a camera that has audio capabilities. This type of configuration allows the homeowner to control the volume within the zone from any of the wall switches associated with it without carrying around a remote control everywhere.

Because the volume controls and speakers do not need to be connected, when the owner wants to change the location of an outlet or speaker, only one cable will need to be changed. This makes moves, adds, and changes (MACs) much easier and more cost-efficient.

WHAT YOU WILL DO

▶   Connect the digital volume control interface to the receiver.

▶   Connect the speaker distribution module to the digital volume control interface.

▶   Connect the in-wall speakers to the speaker distribution module.

▶   Connect the digital volume control to the digital volume control interface.

▶   Power the digital volume control interface.

▶   Test speaker connections.

WHAT YOU WILL NEED

▶   Receiver

▶   Digital volume control interface

▶   Digital volume controller

▶   Speaker distribution module

▶   DC power distribution module

▶   Speaker wire

▶   Category 5e cable

▶   Cable stripper

▶   Electrician's scissors

▶   Four green Phoenix connectors

▶   One gray Phoenix connector

- Two Category 5e plugs
- Screwdrivers
- Category 5e plug crimp tool

OPTIONAL ITEMS

- Digital volume control interface instructions
- DC power distribution module instructions
- Speaker distribution module instructions
- Digital volume controller instructions
- Leviton in-wall speaker instructions

Setting Up

1. If it is not done already, install the digital volume control interface, DC power distribution module, and speaker distribution module into one of the Structured Media Centers (SMCs).

2. Install the digital volume controller if there is not one already located on the lab wall.

3. Check to be sure that the in-wall speakers are installed in the lab wall.

WARNING: Make sure that the receiver is NOT plugged in before beginning this lab.

Part 1: Connecting the Digital Volume Control Interface

1. Roughly measure the distance between the digital volume control interface and the digital volume controller.

2. Cut a piece of Category 5e cable to this length with an additional four or five inches of cable for termination. Terminate both ends of the cable, following the T568A color scheme at both ends.

3. Plug one end of the cable into the digital volume controller and the other end into the digital volume control interface.

   This cable provides the connection between the wall switch and the interface. This cable sends signals only to the volume control interface, not to the audio equipment or speakers.

4. Roughly measure the distance between the digital volume control interface and the stereo receiver.

5. Cut two pieces of speaker wire to this length with an additional four or five inches of wire for termination.

6. Split each end of each wire about two inches and use the small notch on the electrician's scissors (snips) to strip one-quarter of an inch of insulation from both of the conductors at both ends of both cables.

7. Connect one cable to the receiver at the left speaker terminals, being careful to connect the red terminal (+) to the conductor in the speaker wire that has a white stripe or dotted white line on the jacket.

8. Connect the other conductor to the black terminal (–).

9. Connect the other speaker wire to the right speaker terminals. Connect the other end of the speaker wires to one of the Phoenix connectors, observing the left and right connections and the polarity (+ and –) of each channel.

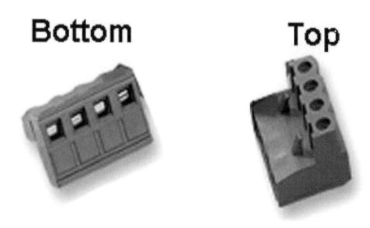

Bottom   **Top**

Figure 5-10-1 Phoenix connector (bottom and top views). The circuit board is connected through the holes at the top of the Phoenix connector. The speaker wires are inserted into the rectangular holes in the bottom of the Phoenix connector.

Refer to Figures 5-10-1 and 5-10-2 when connecting the speaker wires to the Phoenix connector.

After the speaker wire from the amplifier has been connected to the Phoenix connector, plug the connector into the digital volume control interface. This will supply the digital volume control interface with the signals from the left and right channels of the receiver.

10. Plug the connector into the interface marked "AMP."

Part 2: Connecting the Speaker Distribution Module

1. Roughly measure the distance between the digital volume control interface and the speaker distribution module.

2. Cut two pieces of speaker wire to this length with an additional four or five inches of cable for termination.

3. Terminate both ends of the cable with a Phoenix connector, making sure to follow polarity and connecting each conductor to the same connector terminal on the Phoenix connectors. Refer to Figure 5-10-2 when making these connections.

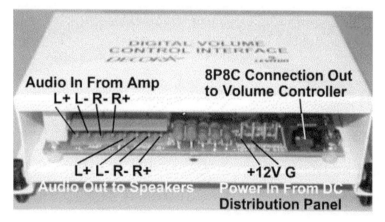

Figure 5-10-2 Connections for the digital volume control interface

4. Now that the speaker wires have been terminated, plug one end into the speaker distribution module indicated by "From Amp."

5. Plug the other end into the digital volume control interface indicated by "SPKR."

This connection will place the audio signal coming from the digital volume control interface onto every connection on the speaker distribution module, allowing multiple pairs of speakers to be controlled from as many as three digital volume controllers.

Part 3: Connecting the Speakers

1. Roughly measure the distance between the digital volume control interface and the left speaker.

2. Roughly measure the distance between the digital volume control interface and the right speaker.

3. Cut two pieces of speaker wire to these lengths with an additional four or five inches of cable for termination.

4. Terminate both conductors of both speaker wires to the same Phoenix connector, making sure to follow polarity and connecting each conductor to the same connector terminal on the Phoenix connectors.

   Refer to Figure 5-10-3 when making these connections.

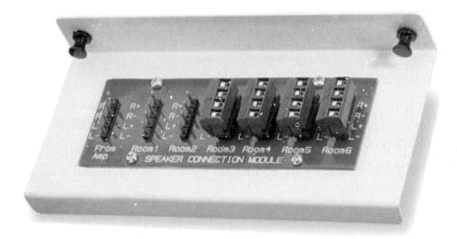

Figure 5-10-3 Speaker connection module

5. Terminate the left speaker wire to the left speaker by loosening the red terminal, inserting the + conductor through the hole in the post, and tightening the terminal.

6. Repeat for the black terminal (–).

7. Repeat for the right speaker.

8. Plug the Phoenix connector with the speakers connected onto the speaker distribution panel in any of the six locations.

Part 4: Powering the Digital Volume Control Interface

1. Roughly measure the distance between the digital volume control interface and the DC power distribution module.

2. Cut a piece of Category 5e cable to this length with an additional four or five inches of cable for termination.

3. Strip about one inch of jacket from each end of the cable.

4. Snip off the orange and blue conductor pairs at the jacket end. These conductors will not be used.

5. Untwist the green pair, strip one-quarter inch of insulation from the ends of the conductors, and twist the copper from the pairs back together.

6. Repeat this for the other end of the cable.

7. Strip and twist the brown pair in the same manner.

This process creates two conductors at each end of the cable. After the cable has been prepared, the connectors can be installed:

1. Attach a green Phoenix connector to one end and the gray Phoenix connector to the other end by using the brown pair for the + and the green pair for the −.

   Make sure that when the connector is installed on the DC power distribution board and the digital volume control interface, the + and − connections line up to the printing on the circuit boards.

2. Plug the gray Phoenix connector end of the cable to the digital volume control interface, indicated by "12V" and "G" on the panel.

3. Plug the other end of the cable to any of the DC connections on the DC power distribution panel.

Part 5: Reviewing the Wiring

1. Check to make sure that each speaker wire from the amplifier—through the digital volume control interface, through the speaker distribution panel, and to each speaker—has the striped conductor to every + or red connection and that the left and right outputs of the receiver reach the correct speakers.

2. Double-check Part 4 to make sure that the electrical wiring is correct. Brown goes to + and green goes to −, both on the DC power distribution panel and on the digital volume control interface.

If you have any doubts about the polarity or positioning of any of the wires or plugs, ask your instructor for assistance.

Part 6: Testing the Installation
After you have checked all of the connections to the speakers and power, you will be ready to test the audio distribution of the hardware.

1. Turn the volume on the receiver to its lowest setting (counterclockwise).

2. Plug in the receiver's power to an AC outlet.

3. Select AM/FM as the source on the receiver and tune in one of the local stations in your area.

4. Make sure that the DC power distribution panel's operation LED is lit (the green light on the panel).

5. Press the top of the volume control switch on the wall until the top LED is lit (this is the maximum output of the volume controller).

6. *Slowly* turn up the volume on the receiver until the source can be heard at a normal listening level.

   Were you able to hear any audio? _____

7. If you could not hear any audio, unplug the receiver and repeat steps 5 and 6. If you still cannot hear any audio, ask your instructor for assistance.

8. Press the bottom of the volume control switch to lower the volume.

   Were you able to lower the audio? _____

9. If you could not lower the audio, unplug the receiver and repeat steps 5 and 6. If you still cannot lower the audio level, ask your instructor for assistance.

10. Press the center of the switch to mute the volume.

    Did the volume mute? _____

11. If the volume was not muted, try pressing the center of the switch again.

12. Press the center of the switch again to unmute the audio signal.

13. If there is a balance control on the receiver, turn the dial all the way to the left.

    What do you hear?

    _____

14. If there is no balance control on the receiver, disconnect the right speaker wire from the red terminal on the receiver.

    What do you hear?

    _____

15. If the speaker has been disconnected, reconnect it to the receiver in preparation for the next lab.

Conclusions

In order to use digital volume distribution, many components need to be installed and configured. Even though there are many devices in use to distribute the audio and provide the capability to control it from several different locations, this method of distribution and control provides quality, convenience, and flexibility when installing a system in the home.

Speaker Loads

INTRODUCTION

The placement of loudspeakers has a great deal to do with the load presented back to an amplifier. This is because speakers are impedances and as such, they obey the rules of series and parallel circuits. Connecting speakers in parallel can reduce the overall impedance. If the impedance is too low, the amplifier could be damaged. Connecting in series can raise the impedance to the point that an amplifier is ineffective.

In a whole house audio installation, the solution is to balance the series and parallel portions of circuits to center the impedance at a value that the amplifier is designed for. This value can be calculated, but the mathematics can be inconvenient and the temptation is to shortcut the process and install the speakers by trial and error. Unfortunately, when dealing with speakers hidden behind the walls, it may be difficult to fix errors.

This lab demonstrates the effects of series and parallel circuit impedances to aid in understanding the speaker-placement process.

WHAT YOU WILL DO

▶   Understand the interaction of speaker impedances in multiroom speaker arrangements.

▶   Learn how to identify harmful impedance combinations before connecting amplifiers to them.

WHAT YOU WILL NEED

▶   Digital Multimeter

▶   12 small jumper cables with alligator clips

▶   8-ohm resistors (to simulate speaker impedance)

Part 1: Preparing the Multimeter

Set your multimeter to ohms. If your meter has both alligator clips and probes, use the clips for this exercise. If your multimeter has an adjustable range, set it to measure in ohms or tenths of ohms, not kilo-ohms or megohms. Depending on the brand and type, you may need to "zero" the meter by touching the probes tightly together and adjusting for a zero reading. Figure 5-11-1 shows a Multimeter.

Figure 5-11-1 Multimeter

Part 2: Examining Resistances

Depending on a number of factors, the measured value of a resistor can vary from its stated value by up to the percentage of tolerance. For instance, a 10-ohm, 10% resistor could have a resistance value of 10 ohms, plus or minus the tolerance (one ohm). Such a resistor could measure from 9 to 11 ohms (10 ohms, plus or minus up to 10% of that value, which is one ohm, hence (10 – 1 to 10 + 1 ohms, or 9 to 11 ohms).

This means that the 8.2-ohm, 5% resistors used in this lab can measure anywhere between 8.2 + 0.41 (5% of 8.2) ohms and 8.2 minus 0.41 ohms, or 7.79 ohms to 8.61 ohms. Keep this in mind as you complete the following exercises.

1. Measure any three resistors and record their values in the spaces below:

    Resistor 1   Resistor 2   Resistor 3

    _____   _____   _____

2. Does the value of each resistor fall within the range 8.2 ohms plus or minus 5% (between 7.79 and 8.41 ohms)? _____ (yes or no) If you encounter a resistor that is outside of that range, double-check your measurement and then set the resistor aside, get a new one, and repeat the Part.

3. Sum together the three values obtained in step 1 and write the answer here. _____

4. Divide this number by 3 (the number of resistors used) to obtain an average value. Is the average within the tolerance of the resistors? (Is it between 7.79 and 8.41 ohms?) _____ (yes or no) Is it closer than the individual values obtained in step 1? _____ (yes or no)

Keep in mind that it is often the case in electronics that individual measurements may vary, but on average they tend to fall into line.

Part 3: Examining Series Resistances

1. Take the three resistors you measured in Part 2 and connect them end to end using the jumpers with alligator clips (see Figure 5-11-2). Connecting electrical loads such as resistors, light bulbs, or loudspeakers end-to-end in this manner is called connecting them in "series."

2. Measure the resistance between points A and B. Write down the value you obtain. _____ Is this value equal or close to the sum of the three resistances that you obtained in step 3 of Part 2? _____ (yes/no)

In series, resistances add together. The total series resistance is equal to the sum of all the individual resistances in the series. Generally speaking, it is the same with inductive loads such as loudspeakers.

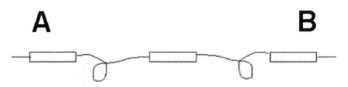

Figure 5-11-2 Resistors connected serially with alligator clips

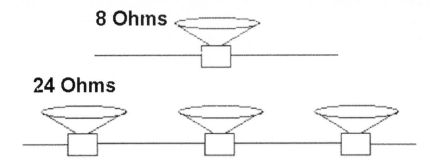

8 Ohms

24 Ohms

Figure 5-11-3 The total resistance is equal to the sum of the individual resistances. In this example, the resistance for one speaker is equal to 8 ohms. The resistance of three speakers is 24 ohms (8 + 8 + 8 = 24).

Part 4: Examining Parallel Resistances

When similar resistive and inductive loads (such as loudspeakers) combine not in a series, but side by side (in parallel), a different result occurs. Research this by creating each of the following resistor configurations. Measure the resistance from point A to point B for each configuration. Write the results in the space below each configuration shown in Figure 5-11-4.

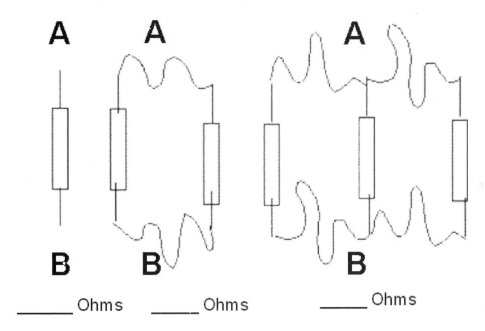

A   A   A

B   B   B

_____ Ohms   _____ Ohms   _____ Ohms

Figure 5-11-4 Resistance measurements

For all configurations, the result should align with the following:

1 resistor =	Resistance of 1 resistor	1 R
2 resistors =	Resistance of ½ resistor	1/2 R
3 resistors =	Resistance of ⅓ resistor	1/3 R

If a second resistor is added next to the first, it provides a second identical path, effectively halving the resistance. A third resistor splits the load three ways, cutting the total resistance to one-third. A fourth cuts total resistance to one quarter, and so on (see Figure 5-11-5).

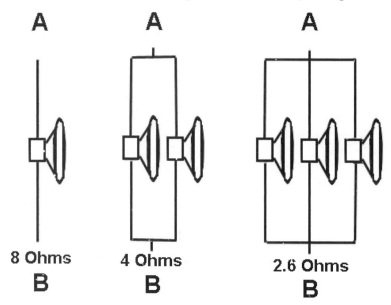

Figure 5-11-5 Resistors reduce the resistance exponentially.

Part 5: Applying Speaker Loads to Residences

If there are too many loads in a series, the amplifier cannot couple adequate power to the speakers due to the impedance mismatch. If there are too many loads in parallel, the resulting impedance will be so low that it may damage the amplifier. There are combinations of series and parallel speakers that will allow you to increase the numbers of speakers and still maintain the correct impedance.

See the floor plan of a small apartment, which is shown in Figure 5-11-6. Try to plot a method of driving these four loudspeakers from one amplifier, while maintaining an impedance of 8 ohms. You may use any combination of series and parallel. Simulate your arrangement with the resistors to check your work, if you wish.

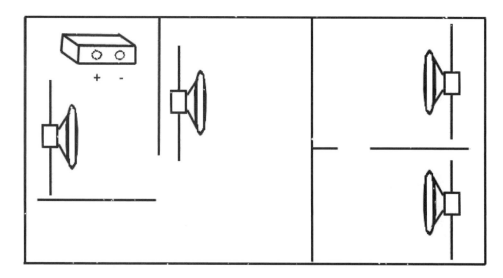

Figure 5-11-6 Floor plan for a small apartment

Figure 5-11-7 provides a hint:

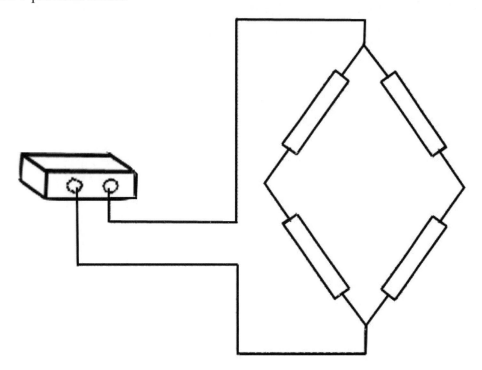

Figure 5-11-7 Ideal placement of four speakers in a small apartment

Even with paying careful attention to impedance distribution, there is a limit to how many speakers can be driven from the same amplifier. Also, arrangements of multiple speakers on a single circuit through a house can be prone to trouble from a number of sources.

Conclusion
This lab demonstrated the effects of series and parallel circuit impedances to aid in understanding the speaker-placement process.

Infrared Systems

INTRODUCTION

One of the many benefits of using a Structured Media Center (SMC) is centralized infrared (IR) remote control. Because all the audio and video equipment cabling runs through a centralized location, this location becomes the hub to which all of the control systems are connected. The homeowner is able to remotely control IR-equipped devices from multiple locations.

Multiple targets and emitters are routed through an enclosure with Category 5e cable, and connected using a 1X9 bridged telephone module. From any room in which a target is located, signals from any IR-emitting device, such as a television remote control, are sent to the same module to be distributed to each device that can be controlled. Emitters located on each device repeat the signal into the IR sensors located in them. Each remotely controlled device will pick up only signals that are meant for it and ignore signals meant for other devices.

Centrally located DC power distribution modules power the emitters and receivers through the same cabling used to distribute the IR signals. This results in less cabling that needs to be installed to control the devices.

WHAT YOU WILL DO

▶ Distribute power and infrared signals through a 1X9 bridged telephone module to power infrared emitters and receivers simultaneously.

▶ Use CAT 5E cable to connect infrared targets to emitters.

▶ Learn how to distribute infrared signals to capable devices.

▶ Control the infrared receiver using the infrared target.

WHAT YOU WILL NEED

▶ Infrared-capable stereo receiver

▶ 1X9 bridged telephone module

▶ Infrared target

▶ Infrared emitter

▶ CAT 5e cable

▶ 110 punch tool

▶ Snips

▶ Cable stripper

▶ DC distribution module

▶ CAT 5e jack

OPTIONAL ITEMS

‣ URL: http://www.ustr.net/infrared/

‣ URL: http://www.howstuffworks.com/inside-rc.htm

Setting Up

1. If it has not been done yet, install the 1X9 bridged telephone module and DC distribution module in one of the SMCs.

2. Also, make sure that the infrared emitter has been attached to the receiver on the sensor (this is usually indicated on the unit). This lab should be done following the audio distribution lab to ensure that there is a usable source and distribution method for it.

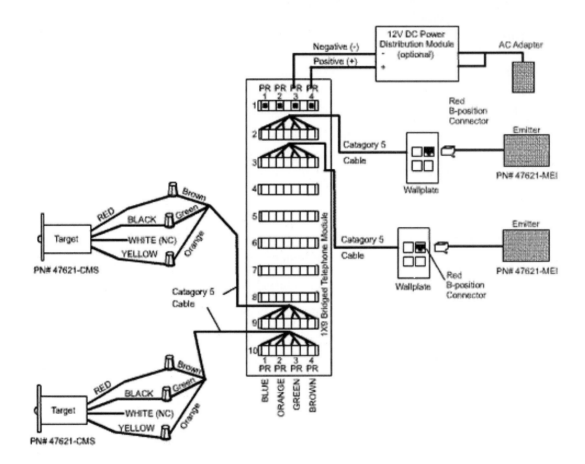

Figure 5-12-1 Schematic diagram showing how infrared targets, wallplates, and DC power distribution module are connected to the telephone module

Part 1: Connecting the Jack

For all the steps in this part, please refer to Figure 5-12-2.

1. Run a length of Category 5e cable from the 1X9 bridged telephone module to an outlet equipped with a module that can accept multiple jacks (see Figure 5-12-2).

2. Terminate one end on the 1X9 module by following the color code indicated on the 110-connections on the module.

3. Punch down the wires on any of the IDC terminals except the top one in the following order:

- ▶ White/blue, blue
- ▶ White/orange, orange
- ▶ White/green, green
- ▶ White/brown, brown

This is the standard order of wire pairs when terminating on a connection block.

How does this method of termination differ from the T568A wiring schematic?

_____

How does this method of termination differ from the T568B wiring schematic?

_____

4. Terminate the other end of the cable to one of the Category 5e jacks and insert it into the outlet module.

5. Plug the cable from the infrared emitter from the receiver into the jack.

Note: As a good rule of thumb, a red jack is normally chosen in this type of situation because 12 VDC will be supplied to this jack. A red jack usually means that power is being supplied to it.

This completes the cabling that will send infrared signals to the infrared-capable device from the infrared distribution panel (the 1X9 distribution panel). Even though this is a telephone line distribution panel, it can be used to distribute many types of signals and even voltage in some cases.

Part 2: Connecting the Target

1. Run a length of Category 5e cable from the bridged telephone module to the location of the infrared target.

2. Terminate one end of the cable to any of the empty IDC connection terminals on the 1X9 module except the top row. Follow the same color scheme as that of the infrared emitter installed on the panel in Part 1.

3. At the other end of the cable, snip off the blue pair and the three white conductors of the other pairs.

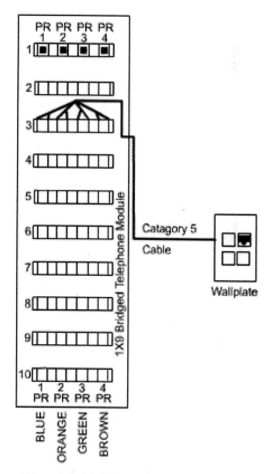

Figure 5-12-2 Running Cat 5 from a wallplate to the telephone module

4. Connect the remaining three conductors to the infrared target, as shown in Figure 5-12-3.

The white wire from the target will not be used.

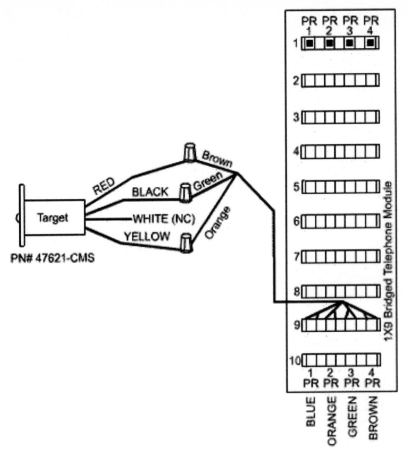

Figure 5-12-3 Connecting the infrared target to the telephone module

Part 3: Connecting the Power

Before connecting any of the power wires, check to make sure that all the previous wiring has been correctly terminated. Ask your instructor to check the connections. This step is very important because if the power wires go to the incorrect connections, the target or emitter could be rendered inoperable.

1. Run a length of Category 5e cable from the DC power distribution module to the 1X9 distribution panel as shown in Figure 5-12-4.

2. Strip the cable and cut off all the conductors except the brown and the green at both ends. These two conductors will supply power to the infrared distribution module.

3. Terminate one end of the cable with a Phoenix connector.

    To which terminal of the Phoenix connector is the brown wire terminated: one of the two on the ends of the connector or one of the two in the middle?

    _____

    To which terminal of the Phoenix connector is the green wire terminated: one of the two on the ends of the connector or one of the two in the middle?

    _____

    Why are these wires terminated there?

    _____

    _____

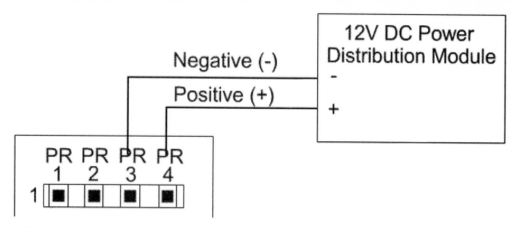

Figure 5-12-4 Running cable between the DC power distribution module and the telephone module

4. Terminate the other end of the cable to the first row of the 1X9 distribution panel.

5. The green wire will be terminated on the sixth position.

6. The brown wire will be terminated on the eighth position.

7. Make sure to double-check this connection BEFORE plugging the Phoenix connector into the DC power distribution module.

   Have your instructor check these connections as well.

   At what position is the positive wire terminated? _____

   At what position is the negative wire terminated? _____

8. After you have double-checked all the connections, plug the Phoenix connector into one of the empty positions on the DC power distribution module.

Part 4: Testing the Infrared Distribution System

In order to test the infrared distribution, the remote control must activate the receiver using only the target that was installed, not the sensor on the front of the device to be controlled.

1. Have one lab partner cover the sensor portion of the receiver (where the infrared emitter is located) with a hand.

2. Have another partner cover the infrared target with a hand.

3. Point the remote control at the receiver and press the power button to make sure that the unit does not respond. If the unit does respond, have each partner check to make sure the units they are covering are completely covered.

4. After the receiver stops responding, ask the lab partner who has been covering the installed target to move away from the target.

5. Point the remote control at the target and try to turn the receiver on. You should see the target light up with a green light inside and the receiver should respond to any requests made by the remote control.

6. If the receiver does not respond to any requests, unplug the Phoenix connector from the DC power distribution panel and check all the wiring against the diagram on the last page of the lab or trace your steps and the wiring through the entire lab.

Conclusion

You have now learned how to distribute infrared signals from targets to emitters.

HAI Control Box Setup

INTRODUCTION

The OmniLT is an automation controller with integrated security for homes and small businesses. The system provides enhanced comfort, safety, convenience, and energy savings by coordinating lighting, heating and air, security, scenes, and messaging based on lifestyles and schedules.

OmniLT comes with several standard modes, such as Day, Night, Away, and Vacation, and can accept customized scenes such as "Good night," "Good morning," or "Entertainment," which set temperatures, lights, and security to the desired levels with one touch. Security and temperature sensors can be used to adjust lights, appliances, and thermostats; monitor activity; and track events. The home can be controlled and programmed by the dealer or the homeowner, on-site or from a remote location. Easy-to-read LCD consoles show status and allow control and scheduling of lighting, security, temperatures, and accessories.

OmniLT has a built-in serial interface for connection to the Internet via HAI Web-Link II, personal computers, and options such as touchscreens, voice recognition, and home theater controls. OmniLT features telephone accessibility from within the home or a remote location with clear voice menus for convenience and simplicity. The built-in digital communicator reports alarm events to a central station and can dial up to eight additional phone numbers chosen by the home owner for voice notification.

The Model 21A00-11 Leviton version of the OmniLT includes specially designed brackets for mounting in the Leviton Structured Media Center Enclosures, which serve as a home networking and entertainment hub.

WHAT YOU WILL DO

- ▸ Connect the controller box to the 16 VAC transformer.
- ▸ Connect the console to the controller box.
- ▸ Connect the X-10 module to the controller box.
- ▸ Connect *one* lead to the battery.

WHAT YOU WILL NEED

- ▸ HAI controller box
- ▸ HAI console
- ▸ Category 5e cable
- ▸ Screwdrivers
- ▸ Snips
- ▸ Wire stripper
- ▸ HAI control box battery
- ▸ Model 10A09 Dual X-10 Transmitter Kit
- ▸ 110 VAC to 16 VAC transformer

Setting Up

1. Place the battery inside the HAI housing. This is used as a backup for the system if there should be a failure with the AC connection.

2. Roughly measure the distance from the nearest AC outlet to the HAI housing.

3. Cut a length of Category 5e cable equal to this distance with an additional eight to ten inches allowed for termination. If this is the first time the console will be connected to the HAI board, roughly measure the distance from the HAI console to the controller box and cut a length of cable equal to this distance with an additional eight to ten inches allowed for termination.

All subsequent connections of the console to the board will only be at the board side. The console will already be connected to the Category 5e cable.

Remember that at no point in this lab will the transformer be plugged into any outlet, nor will the battery be completely connected to the controller board.

Part 1: Creating a Power Cord

1. Strip approximately two inches of jacket from the ends of the Category 5e cable.

2. Cut off the blue and orange pairs at the end of the jacket. These pairs will not be used.

3. Use the snips to strip about one-quarter inch from each of the wires in the remaining pairs.

4. Twist the copper of the brown pair together and repeat for the green pair at each end of the cable. This will create two connections at each end of the cable. Because this is a low-voltage connection, these will suffice as the ungrounded and grounded conductors (hot and neutral) of the AC connection between the HAI circuit board and the power transformer.

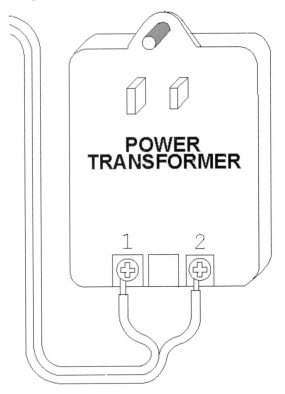

Part 2: Connecting the Power Cord

*Make sure that the transformer is **not** plugged in.*

1. As shown in Figure 6-1-1, the transformer has two screws used as terminals for the cord connections. Loosen the screws on the AC transformer.

2. Insert the brown pair under the terminal marked 1 on the transformer and tighten the screw.

Figure 6-1-1 Connecting the power cord to the transformer

3. Insert the green pair under the terminal marked 2 on the transformer and tighten the screw.

Loosen the second and third terminals from the far left of the HAI box terminals (see Figure 6-1-2).

4. Route the cord into the HAI box through the hole in the back of the enclosure. Place the *BROWN* pair in the first hole of the terminal pair marked "16.5 VAC INPUT" and tighten the screw. Place the *GREEN* pair in the second hole of the terminal pair marked 16.5 VAC INPUT and tighten the screw.

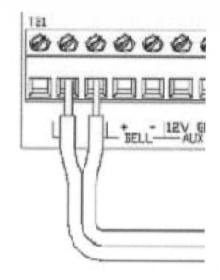

Connect the BLACK battery wire to the minus (–) terminal on the battery. *DO NOT* connect the red wire at this time. *DO NOT* reverse the connections; the battery fuse will blow. Note that the unit will *NOT START* using the battery alone.

DO NOT *plug in the transformer at this time.*

Figure 6-1-2 Connecting the power cord to the HAI box

Part 3: Connecting the X-10 Interface

The HAI X-10 interface module sends commands to X-10 enabled devices from the HAI controller board. In order to send these commands, the HAI board must interface with the AC house line using a special X-10 module.

Use the four-conductor modular telephone cable to connect the jack on the controller board. This is the jack on the right of the controller board designated RJ11.

Route the cable around the controller board and out through the hole in the back of the housing. Plug this cable into the X-10 interface module.

Part 4: Connecting the Console

Connecting the Category 5e cable to the supplied console cable using the supplied connectors performs a one-time connection of the cable to the console. This is done only one time by the instructor to avoid running out of wire length on the supplied connector.

Route the cable from the console to the control box through the hole in the back of the enclosure, as shown in Figure 6-1-3.

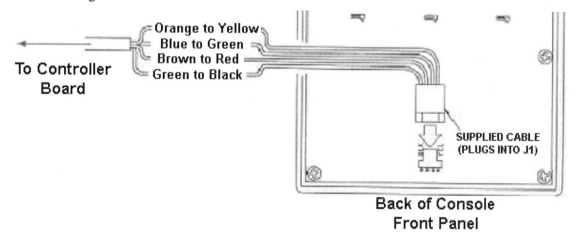

Figure 6-1-3 Connecting the console

Strip about two inches of jacket from the end of the cable. Strip about one-quarter inch of insulation from the ends of each of the eight conductors. Twist the copper of each pair together to form four conductors.

Connect the conductors to the controller board by loosening the four screws in the section of the controller board terminals marked **CONSOLE** enough to slide the conductors into the holes and then tightening them securely (see Figure 6-1-4).

The **BROWN** pair will be connected to the **RED** terminal.

The **GREEN** pair will be connected to the **BLK** terminal.

The **ORANGE** pair will be connected to the **YEL** terminal.

The **BLUE** pair will be connected to the **GRN** terminal.

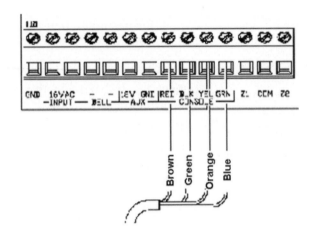

Figure 6-1-4 The conductors securely inserted into the console terminals

Double-check all connections to make sure that they are correct and the terminals are secured. Figure 6-1-5 shows the final console connections.

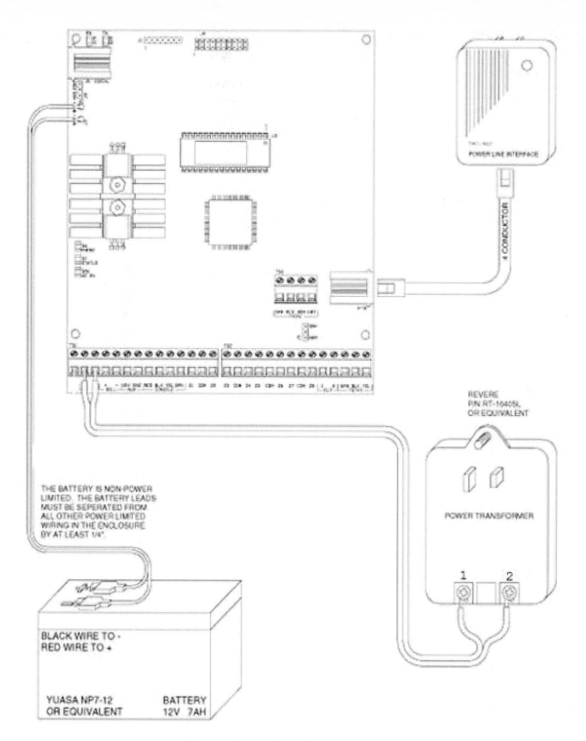

Figure 6-1-5 Console connections

Lab 6-1

Video Sequencer Configuration

INTRODUCTION

This lab demonstrates the capability to sequentially display images and audio from up to four different sources, normally security cameras, on a television. The images can be manually advanced or set to automatically sequence. If the images are set in automatic mode, the time delay between the displayed images is variable.

Prior to doing this lab, you should have completed Lab 5-6, "Video Camera Installation." The cabling and connections from that lab are used in this one.

WHAT YOU WILL DO

▶ Connect two video camera outputs to the video sequencer.

▶ Connect the audio output from the exterior camera to the sequencer.

▶ Display the sequencer output on the television.

▶ Use manual mode and vary the time delay between images in automatic mode.

WHAT YOU WILL NEED

▶ Video sequencer installed and powered in the SMC

▶ Television with composite video-in and audio-in capability

▶ RG-6 coax with camera video output from two security cameras

▶ Audio-out from the exterior camera (jumper cable with white RCA male plug in the SMC)

▶ Two RG-6 coax patch cords with F-to-RCA male adapters on each end

OPTIONAL ITEMS

VCR with RG-6 coax patch cable with F type connectors

Setting Up

1. Locate the television near the SMC containing the video sequencer (see Figure 6-2-1).

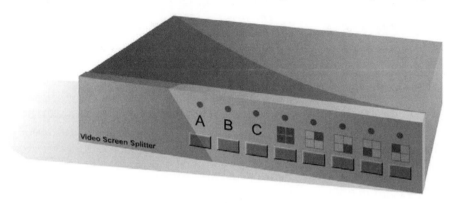

Figure 6-2-1 Video sequencer

2. Make sure that you have two RG-6 coax patch cables with F-to-RCA male adapters long enough to reach from the video sequencer to the inputs on the television.

3. Verify that both cameras are powered on.

Part 1: Connecting Signal Input

1. Connect the coax cable with the output from the Decora interior camera to input #1 on the video sequencer.

2. Connect the coax cable with the output from the exterior camera to input #3 on the video sequencer.

3. Connect the white RCA male audio jumper to the left audio on input #3.

Part 2: Connecting Signal Output

Can audio signals travel over RG-6 coax, which is normally used for video signals?

Audio signals can also use coax, which offers excellent shielding and low resistance. In fact, most of the multicable jumpers (yellow, red, and white) that come with VCRs to connect to the television are made up of three coax cables: one for video and two for stereo audio.

1. Connect one of the RG-6 coax patch cables from the left audio-out on the video sequencer to the audio-in on the television.

2. Connect the other patch cable between the video-out on the video sequencer and the video-in on the television.

Part 3: Testing

1. Turn on the television and select the video-in as the signal source.

2. Turn on the power switch on the video sequencer. The image from one of the cameras should be displayed on the television. If it is the exterior camera, the audio from that camera should come through the speakers.

3. If no camera image is displayed, verify that all of the cables in steps 1 and 2 are connected properly.

Part 4: Operating

1. Use the mode selector on the video sequencer to select manual mode. Notice that there are green LEDs on the front face of the sequencer that indicate which video source is being used.

2. Step through the different input sources and notice the display change on the television.

3. Set the video sequencer to automatic mode. The LEDs on the front panel should indicate that it is stepping from one active input to another.

4. Use the speed control to adjust the step rate from between 1 and 30 seconds.

Why are there no "blank" screens for the unused inputs on the video sequencer?

The sequencer can sense which inputs have an active signal and it will skip over any that do not. This makes it much easier for the installer, who does not have to set up any programming or make any changes to the sequencer when adding or removing cameras.

Part 5: Connecting to a VCR (Optional)

The video and audio signals that are displayed on the television may be recorded on a VCR. To use the VCR:

1. Disconnect the coax patch cables from the audio-in and video-in on the television and connect them to the audio-in and video-in on the VCR.

2. Connect another coax patch cable with F connectors from the VCR antenna-out to the television antenna connection.

3. Set the television to use its tuner and select channel 3 or 4 (this should match the channel-out selection on the back of the VCR).

4. The VCR must be turned on. It may have to be configured to select the audio and video input. The TV/Video mode may also have to be set to video.

You should now see images displayed on the television just as you did before the VCR was put in place. If you insert a videotape and select Record, those images will be recorded on the tape. *Note:* For security-monitoring cameras, time-lapse VCRs are used. They record many more hours than conventional VCRs can.

Conclusions

During this lab, you learned how to configure devices to allow a television to display the output from multiple video cameras and use a video sequencer to automatically display the images for a variable time period. If the optional VCR was used, you demonstrated that those same images and audio could be recorded on videotape.

Do not remove any of the connections you have made because they will be used in a future lab.

HAI Thermostat Connections

INTRODUCTION

The RC-Series is a family of electronic communicating thermostats that can be controlled both locally and remotely. The RC-Series includes models for conventional single-stage heat/cool, heat pump, two-speed heat pump, two-stage conventional, and zone control systems.

The large LCD display shows time, temperature, and operational mode. The controls are easy to use, with simple raise and lower temperature keys, mode (off, heat, cool, auto), and fan keys (on, auto). The Prog key is used for programming when the thermostat's internal scheduling features are used. The Hold key maintains the current settings as long as the hold indicator is on. A remote indicator shows when the thermostat has been set by remote control.

The thermostats feature a multimode communications interface for connection to many types of home and building automation systems. This allows more comprehensive energy and comfort management by setting the thermostats based on occupancy and home or building modes along with traditional scheduling. Two full communications modes are available for connection to automation systems and personal computers. Two additional modes are available for simplified connection to most access control, security, and time clock systems.

When used with HAI's Omni family of automation controllers, full control of RC-Series thermostats is available on remote LCD consoles, on telephones (with voice response), and over the Internet with HAI Web-Link II. The Omni's advanced programming features add comprehensive energy-management capabilities to one or more thermostats. The Omni automatically sets the thermostat's time and outdoor temperature displays.

All models feature advanced logic for temperature control, including "anticipation" for maximum comfort. Models with more than one stage feature HAI's advanced Energy Efficient Control (EEC) to minimize the use of expensive auxiliary heat. The installer can set limits on the heat and cool settings to prevent the thermostat from being set too high or too low. The thermostat prevents short cycling, protecting the system's compressor. A filter reminder tracks the running time of the fan, indicating when it is time to replace the filter.

WHAT YOU WILL DO

▸ Connect the thermostat to the controller board.

▸ Connect the thermostat to the AC transformer.

▸ Connect a temperature sensor to the controller board.

WHAT YOU WILL NEED

▸ Thermostat

▸ Temperature sensor

▸ Snips

▸ Cable stripper

▸ Category 5e cable

▸ Screwdrivers

▸ Thermostat manual

▸ OmniPro manuals

▸ Wire nuts

▸ Label maker

▸ 24 VAC transformer

OPTIONAL ITEMS

URL: http://www.homeauto.com

SETTING UP

1. Roughly measure the distance from the AC transformer to the thermostat.

2. Cut a length of Category 5e equal to this distance, with an additional eight to ten inches allowed for termination.

3. Roughly measure the distance from the temperature sensor to the controller board.

4. Cut a length of Category 5e cable equal to this distance with an additional eight to ten inches allowed for termination.

5. If this is the first time the thermostat will be connected to the HAI board, roughly measure the distance from the HAI console to the controller box and cut a length of Category 5e cable equal to this distance, with an additional eight to ten inches allowed for termination.

All subsequent connections of the thermostat to the board will be only at the board side. The console will already be connected to the Category 5e cable.

Part 1: Creating a Power Cord

1. Strip approximately two inches of jacket from the ends of the Category 5e cable that was cut to length from the thermostat to the AC transformer.

2. Cut off the blue and orange pairs. These conductors will not be used.

3. Use the snips to strip about one-quarter inch from each of the wires.

4. Twist the copper of the BROWN pair together and repeat for the GREEN pair. This will create two connections at each end of the cable. Because this is a low-voltage connection, these will suffice as the ungrounded and grounded (hot and neutral) conductors of the AC connection between the thermostat and the 24 VAC power transformer.

Part 2: Connecting the Power Cord

1. Remove the faceplate of the thermostat by inserting a small screwdriver into the small holes on the bottom of the thermostat housing. This will release the clips that hold the face on the mounting plate.

2. Route the power cable through the back of the thermostat, as shown in Figure 7-1-1.

3. Loosen the top and bottom terminal screws, marked R and C, located on the thermostat mounting plate (see Figure 7-1-2). These two terminals will connect to the AC transformer to supply power to the thermostat. This connection

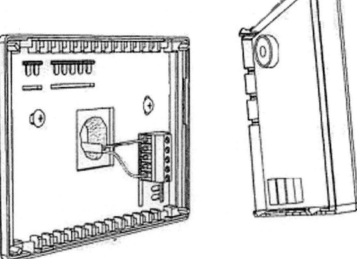

Figure 7-1-1 Routing the power cable through the back of the thermostat

is made in the lab to simulate the power connections from the HVAC unit normally present in the home.

4. Insert the BROWN pair into the top terminal marked R and tighten the terminal.

5. Insert the GREEN pair into the bottom terminal marked C and tighten the terminal.

6. Loosen the left and right terminals on the 24 VAC TRANSFORMER enough to slip the joined conductors under the screws. Connect the BROWN pair to the terminal marked 1. Connect the GREEN pair to the terminal marked 3.

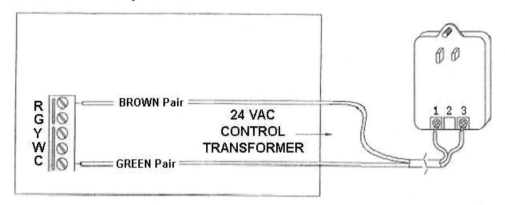

Figure 7-1-2 Connecting the transformer to the thermostat

7. Inspect all of the connections to make sure that they are correct and secure. Ask your instructor to check your work to ensure that the connections are correct.

Part 3: Connecting the Thermostat to the Controller Board

Connecting the yellow, green, red, and black conductors of the supplied thermostat cable to Category 5e cable using the supplied connector performs a one-time connection of the cable to the thermostat. This is done only one time to avoid running out of wire length on the supplied connector. (All future labs will already have this step performed.)

Important: As shown in Figure 7-1-3, when connecting the supplied thermostat cable to the Category 5e cable, be sure to connect *both* the BLACK and RED conductors of the thermostat cable together to the BROWN conductor of the Category 5e cable. The YELLOW wire is connected to the ORANGE conductors and the GREEN wire is connected to the GREEN conductors. The BLUE conductors of the Category 5e are not used in this connection.

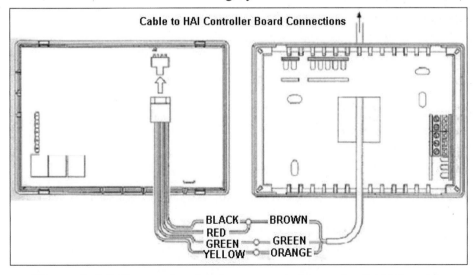

Figure 7-1-3 Connecting the thermostat cable to the CAT 5e cable

HAI Thermostat Connections   129

1. Snap the thermostat faceplate back onto the mounting plate.

2. Route the cable from the thermostat behind the control box and through the hole in the back of the enclosure.

3. Strip about two inches of jacket from the other end of the Category 5e cable. Cut off the blue wire at the end of the jacket. Strip about one-quarter of an inch of insulation from the ends of each of the six conductors (the blue wire is not used). Twist the copper of each pair together to form three conductors.

4. Connect the conductors to the controller board by loosening the three screws in the section of the controller board terminals marked TSTAT; loosen enough to slide the conductors into the holes and then tighten them securely.

   See Figure 7-1-4 for the correct wiring.

   ▶ The BROWN conductors will be connected to the BLK terminal.

   ▶ The GREEN wire will be connected to the GRN terminal

   ▶ The ORANGE conductors will be connected to the YEL terminal.

If more than one thermostat is used in the system, they must be connected in series to the first thermostat.

Use the label maker to label the cable Thermostat 1.

Part 4: Connecting the Remote Temperature Sensor

The Model 14A00 is a temperature sensor that can mount in an outdoor location, usually under an overhang, and sends the outdoor temperature to an OmniLT system. It is coated with a sealant to withstand outdoor moisture. The outdoor temperature can be displayed on the console, spoken over the telephone, or displayed on an HAI communicating thermostat.

Remote temperature sensors can be used to measure temperatures where no thermostat is installed and the homeowner wants to monitor the location. This is useful when the homeowner wants to open or close a skylight or drapes when temperatures rise or fall to a certain extent, turn on or off a fan when the temperature reaches a certain point, or even turn on or off a misting system at a specific outside temperature.

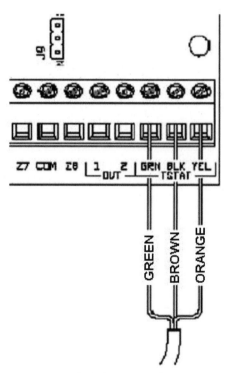

Figure 7-1-4 Connectors connected to the TSTAT terminals

1. Strip about two inches of jacket from one end of the cable that was cut to length for the temperature sensor.

2. Cut the BLUE conductors at the jacket end. These wires will not be used.

3. Strip about one-quarter inch of insulation from the ends of each of the remaining conductors.

4. Twist the copper of each pair together to form three conductors.

5. Connect the BROWN pair to the RED wire, connect the GREEN pair to the BLACK wire, and connect the ORANGE pair to the YELLOW wire using wire nuts.

6. Route the cable to the back of the HAI controller box and through the hole in the back of the enclosure.

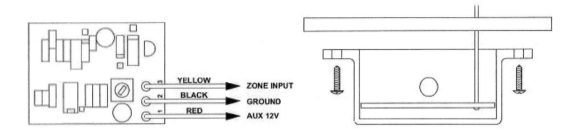

Figure 7-1-5 Connecting the remote temperature sensor

7. The BROWN and GREEN pairs are connected to the controller board at the Auxiliary section of the board to provide the unit with power. The BROWN wire is connected to the terminal marked 12V. The GREEN wire is connected to the terminal next to it, marked GND.

8. The ORANGE wire is the one that reports the temperature back to the controller board. This connection can be to any of the zones on the controller board. For this lab, connect it to the Z3 terminal of the terminal pair. Zone three will be used because it is very close to the Auxiliary power connections, making a "clean" installation.

9. Use the label maker to label the cable Temperature Sensor.

10. Double-check all connections to make sure that they are correct and the terminals are secured.

HAI System Power-Up

INTRODUCTION

Now that the connections from the controller, thermostat, and temperature sensor have been made to the controller board, it is time to power-up the system and make sure that all of the connections have been made correctly.

WHAT YOU WILL DO

▶   Power up the HAI system.

▶   Configure thermostat settings.

WHAT YOU WILL NEED

▶   HAI installation manual

▶   Thermostat installation manual

OPTIONAL ITEMS

URL: http://www.homeauto.com

Setting Up

1.   Carefully review hookups to the zones, grounds, sounders, and consoles.

2.   Disconnect one lead of both the interior and exterior sounders if they are installed.

3.   Make sure that the red positive lead to the battery is disconnected at this time. Make sure that it is not touching anything.

Part 1: Plugging in the Transformers

1.   Plug in the X-10 interface to the AC outlet.

2.   Plug in the 24 VAC power transformer first.

3.   Plug in the 16 VAC power transformer second.

The AC ON LED should illuminate on the left side of the controller board. If it does not illuminate, unplug the transformer immediately and check all connections carefully. Ask your instructor for help if necessary.

Within one minute, the STATUS LED should begin blinking one time per second. This indicates that the processor and software are working.

The PHONE LED should be on.

4.   Unplug the 16 VAC power transformer.

5.   Connect the red battery wire to the positive battery terminal. The system should *not* start.

6.   Plug in the 16 VAC power transformer. The system should start.

7. Unplug the 16 VAC power transformer to make sure that the battery will back up the system. The system should continue to run on the battery (the STATUS LED will continue to flash).

8. Plug in the transformer again to resume AC power.

Part 2: Checking Out the Console

The console should be operating. Figure 7-2-1 shows the console.

Figure 7-2-1 HAI console

1. Press the * (asterisk) key to silence the trouble beeper if it is beeping.

2. Press OFF, 1, 1, 1, 1 if the alarm is tripped.

3. If the console is not operating correctly, unplug the 16 VAC transformer and carefully check all the connections. Ask the instructor to help you if necessary.

The first function that will need to be performed is the clearing of the system memory from the previous class. This step is very important to perform because it will return the controller to its factory default settings. First the EEPROM must be reset, and then the system RAM must be reset.

4. To access the installer setup mode, first press the 9 key.

5. Enter the installer code 1, 1, 1, 1.

   This will access the setup menu.

6. Press the # key to access the Installer setup menu.

7. Press the 6 key to access the misc. configuration menu.

8. Press the down arrow key until you find the Reset System EEPROM option.

9. Press the 1 key to indicate "yes."

10. Press the # key to accept this option.

11. Press the * (asterisk) key repeatedly to return to the main screen.

12. To access the installer setup mode, first press the 9 key.

13. Enter the installer code 1, 1, 1, 1.

    This will access the setup menu.

14. Press the # key to access the Installer setup menu.

15. Press the 6 key to access the misc. configuration menu.

16. Press the down arrow key until you find the Reset System RAM option.

17. Press the 1 key to indicate "yes."

18. Press the # key to accept this option.

19. Press the * (asterisk) key repeatedly to return to the main screen.

Because there is no telephone line connected to the equipment, you must disable the controller board from detecting the status of the phone.

20. To access the installer setup mode, first press the 9 key.

21. Enter the installer code 1, 1, 1, 1.

    This will access the setup menu.

22. Press the # key to access the Installer setup menu.

23. Press the 6 key to access the misc. configuration menu.

24. Press the down arrow key until you find the Dead Line Detect option.

25. Press the 0 key. Press the # key to disable the detection of the telephone system.

26. Press the * (asterisk) key repeatedly to return to the main screen.

The bottom line of the display should read SYSTEM OK.

27. If there are any trouble indications that occurred during installation, press * (asterisk) key to acknowledge them and silence the beeper.

28. Set the time and date by pressing the 9 button. Enter the Master Code (1, 1, 1, 1); then press the 2 button.

29. Enter the time on the keypad and then the date. Time and date are entered as four and six numbers, respectively.

30. Press the * asterisk key to exit the menus.

The console should now show the time and date on the top line and SYSTEM OK on the bottom line.

Part 3: Checking Out the Thermostat

Observe the display on the thermostat. The current temperature reading should show on the display. Figure 7-2-2 shows the HAI thermostat.

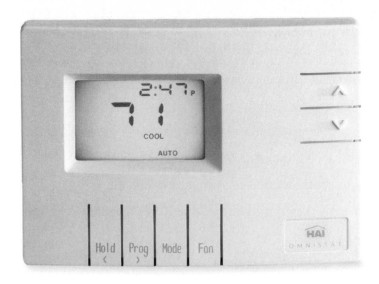

Figure 7-2-2 HAI thermostat

1. Press the FAN key. FAN should be displayed on the screen. Press the FAN key again. FAN should disappear from the screen.

2. Set the mode to HEAT. Use the up arrow key to raise the desired temperature setting above the current temperature. After a few seconds, you will hear a click, indicating that the unit has sent a signal to the HVAC unit to turn on the heat. Set the mode to OFF. Ensure that the unit displays OFF.

3. Set the mode to COOL. Use the down arrow key to lower the desired temperature setting below the current temperature. After a few seconds, you will hear a click, indicating that the unit has sent a signal to the HVAC unit to turn on the cooler. Set the mode to OFF. Ensure that the unit displays OFF.

4. If the thermostat or system does not perform as stated previously, recheck all wiring carefully. Ask the instructor for help if you need it.

Part 4: Setting Up Thermostat Values

You need to know the following about Installer Setup mode, which is used to configure the general operating parameters of the thermostat:

▸ The small digits on the top of the display show the item number being set.

▸ The large blinking digits in the center of the display are the value of the item number.

▸ The PROG (>) key advances to the next item.

▸ The HOLD (<) key returns to the previous item.

▸ The arrow keys (^ – ⱽ) change the value of each item.

▸ *Never set the values to anything other than the specified range for each item.*

▸ To exit Setup mode, press the FAN key.

To set up thermostat values, do the following:

1. To enter Installer Setup mode, set MODE to OFF.

2. After 10 seconds, press the PROG key three times (day will flash) and then press the FAN key. Note the following settings:

- ▶ Item 00: Address. Only one thermostat will be used, so the default value of 1 will be fine.

- ▶ Item 01: Communications Mode. The HAI system will use Mode 1, which is the default setting and will not need to be adjusted.

- ▶ Item 02: System Options. The default setting of 0 will be fine.

- ▶ Item 03: Display Options. The default setting of 1 will need to be changed to 5. Use the down arrow key to change this and press # to accept the change.

- ▶ Item 04: Calibration Offset. The default setting of 30 will be fine.

- ▶ Item 05: Cool Setpoint Limit. The default setting of 51 will be fine.

- ▶ Item 06: Heat Setpoint Limit. The default setting of 91 will be fine.

- ▶ Items 07 and 08 are not used.

- ▶ Item 09: Cooling Anticipator. The default setting of 4 will be fine.

- ▶ Item 10: Heating Anticipator. The default setting of 4 will be fine.

- ▶ Item 11: Cooling Minimum Time. The default setting of 8 will be fine.

- ▶ Item 12: Heating Minimum Time. The default setting of 8 will be fine.

- ▶ Item 13 is not used.

- ▶ Item 14: Clock Adjust. The default setting of 30 will be fine.

3. Press the FAN key to exit Installer Setup mode.

More information about these settings can be found in the thermostat installation manual.

Troubleshooting tips can also be found in the installation manual.

If you encounter any problems while setting the thermostat, ask the instructor to help you.

Programming X-10 Lighting Controls

INTRODUCTION

Leviton universal dimmers (HCM06) are designed for use with DHC residential power line carrier components and function as remote dimming devices that respond to coded DIM/BRIGHT, ON/OFF, and ALL LIGHTS ON/OFF commands. Commands are delivered through the X-10 power line carrier protocol.

Leviton Controllers (HCCxx) are signal transmitters that work in conjunction with DHC receivers. The wall-mounted DHC transmitter is designed to provide manual control of loads from one or more locations in a home over standard 60 Hz 120 V wiring. The HCCxx transmitters incorporate Leviton's exclusive Intellisense™ Automatic Gain Control (AGC) circuitry, which is a variant of the X-10 power line carrier. Intellisense transmitters are compatible with standard X-10 devices.

WHAT YOU WILL DO

▶   Assign addresses to X-10 receive devices.

▶   Assign an address to the transmitting controller.

WHAT YOU WILL NEED

▶   Small screwdriver

▶   Ballpoint pen

Setting Up

Decora home controls (DHC) work together as a communications network. DHC programmers and controllers are the transmitters in this network. They are installed at a location in the home where convenient for their designated use. All transmitters send a unique coded command signal throughout the home's existing AC wiring, which links the entire network together. DHC switch, fixture, and receptacle modules (which act as receivers in the network) are installed in place of standard switches, outlets, and dimmers at key locations throughout the home.

The system communicates via "preset addresses" set on controllers and receivers (dimmers, switches, and receptacles). An address consists of one letter code from A–P and one number code from 1–16 (see Figure 7-3-1). There are 256 letter-number combinations available for use as system addresses. Each unit must be set (programmed) to allow it to communicate with devices in the home controls system. This is accomplished by setting a controller's transmit code to correspond with the address settings of the receivers to be controlled in the system in which it is installed.

Activity

1. Using a small screwdriver, remove the cover of the seven-scene controller by prying carefully at the sides of the cover (see insert at bottom left in Figure 7-3-1).

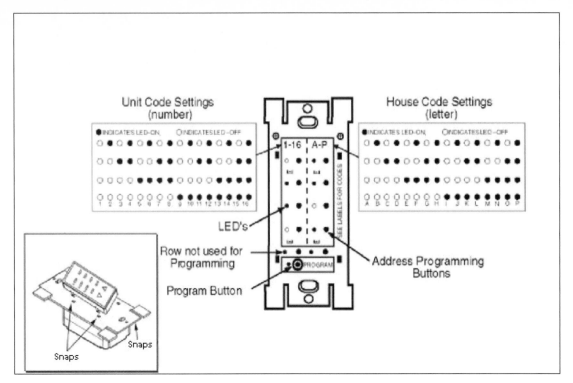

Figure 7-3-1 Scene Controller

2. Identify the programming buttons. The left row of buttons programs the unit code; the right row of buttons programs the house code. The program button is the lone button at the bottom of the unit.

3. Press and release the program button.

LEDs should flash; steady lit LEDs identify the unit's program. The default program is A1, which is when no LEDs are lit.

4. Depress the second LED on the unit code column and the third LED on the house code column. This will program the unit to be E3.

5. Depress the program button to set the program into the unit.

6. Depress the program button.

The second LED in the unit code column and the third LED in the house code column should be lit.

7. Depress the buttons associated with these LEDs. The LEDs should extinguish.

This returns the scene controller to A1, which is the original setting.

8. Depress the program button so that the program "sets."

Conclusions

X-10 controllers can be programmed with up to 16 unit codes and house codes that range from A to P. A controller can control up to 16 devices on a single house code. Controllers cannot control devices on different house codes.

Addressing Dimmers

INTRODUCTION

Dimmers and light switches are considered to be receiving devices, receiving X-10 signals from control devices that are transmitters. Receiving devices must be set to the same house code as the transmitting devices. Each receiving device must have a unique unit code.

WHAT YOU WILL DO

▸ Set address codes to dimmers and light switches.

▸ Set lighting scenes on the Leviton scene controller.

WHAT YOU WILL NEED

▸ Table top X-10 controller

▸ Ballpoint pen

Setting Up

▸ Plug the X-10 tabletop controller into an electrical outlet on the lab wall.

▸ Set the house code to A.

For HCM06/10 receivers with one-button programming, there are no code wheels. To set their address, use a small pointed object (very small screwdriver) to depress and hold the program switch until the ON/OFF LED flashes (see Figure 7-4-1).

Part 1: Addressing the First Dimmer

Identify the dimmer nearest the Leviton scene controller.

1. Using a ballpoint pen or similar object, depress the program button on the dimmer, as shown in Figure 7-4-1.

   Wait until the small LED at the bottom flashes.

2. On the tabletop controller, press the number 1 button to on.

   This sends a signal to the dimmer and turns the light on.

3. Press the number 1 button on the tabletop controller to off.

   The light now goes out and the dimmer is set to 1A.

4. Label this dimmer as 1A.

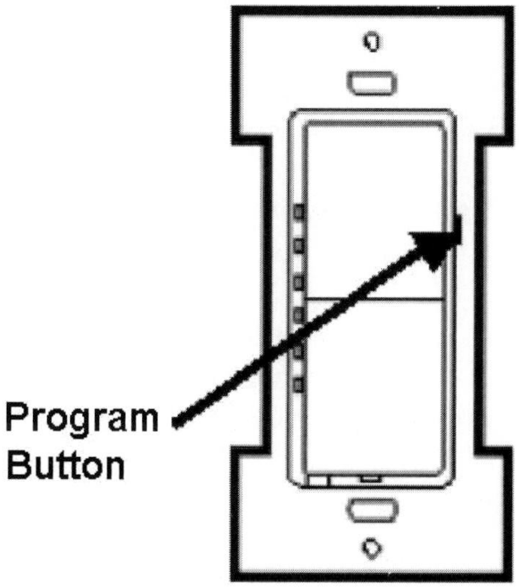

Note: It is acceptable for the controller and one device on the system to have the same address.

Program Button

Figure 7-4-1 HCM06/10 receiver with one-button programming

Part 2: Addressing the Second Dimmer

1. Go to the next dimmer.

2. Follow the steps in Part 1 to program this dimmer to 2A.

3. Press 2 on the tabletop controller.

4. Label this dimmer.

Part 3: Addressing Additional Dimmers

Set address for two more dimmers or switches to 3A and 4A.

Part 4: Preparing the Scene Controller

With all lights off, press and hold the scene 1 button on the scene controller until all LEDs on the controller flash. This may take 5 to 30 seconds.

Part 5: Programming the First Scene

1. Set the light on dimmer A1 to bright.

2. Set the light on dimmer A2 to dim.

3. Press the scene 1 button on the scene controller.

4. Press the off button on the scene controller.

Part 6: Programming the Second Scene

1. Press the scene 2 button on the scene controller and hold it to put it into programming mode.

2. Set the light on dimmer A2 to bright.

3. Set the dimmer on A1 to dim.

4. Press the scene 2, program button.

Part 7: Programming Additional Scenes

Following the procedures in Parts 5 and 6, set scene 3 and scene 4 at your discretion.

Part 8: Testing the Scenes

1. Press scene buttons 1 through 4 on the scene controller.

2. After a scene is pressed, press the off button before going to the next scene button. Each scene should light in the preprogrammed scheme.

HAI Remote Thermostat Control

INTRODUCTION

When used with HAI's Omni family of automation controllers, full control of RC-Series thermostats is available on remote LCD consoles, telephones (with voice response), and over the Internet with HAI Web-Link II. The Omni's advanced programming features add comprehensive energy management capabilities to one or more thermostats. The Omni automatically sets the thermostat's time and the outdoor temperature displays.

WHAT YOU WILL DO

▶ Set the Omni controller board to accept input from the thermostat.

▶ Set the Omni controller board to accept input from the remote temperature sensor.

▶ Send commands to the thermostat from the Omni console.

WHAT YOU WILL NEED

▶ Omni installation manual

▶ Thermostat installation manual

OPTIONAL ITEMS

▶ URL: http://homeauto.com

▶ Omni quick reference command summary

Setting Up

Ensure that the Omni system and connected devices are connected and operating correctly.

Part 1: Setting Up the Thermostat

These setup items are stored permanently in the system, even if the battery and 24 VAC power are disconnected. The default settings are the ones that have been set at the factory. In order to operate and monitor thermostat settings and remote temperature readings, the system must be configured to understand what type of device has been connected to which terminals and zones. Figure 7-5-1 shows the HAI controller.

Figure 7-5-1 HAI controller

First, follow these steps to set up the thermostat:

1. To access the installer setup mode, first press the 9 key.

2. Enter the installer code 1, 1, 1, 1.

   This will access the setup menu.

3. Press the # key to access the Installer setup menu.

4. Press the 5 key to access the temperature configuration menu.

5. Press the down arrow key to move to the thermostat setting type item.

   This item specifies the thermostat type for each thermostat.

   Setting this item will enable the thermostat.

   For each thermostat type, the current setting is shown on the bottom line.

6. To enable or change a thermostat type, press the # key. Use the arrow keys to scroll through the list of thermostat types; then press # to select a new type.

7. Set the thermostat type to AUTO HEAT/COOL by choosing #1.

8. Press the * (asterisk) key repeatedly to exit the installer configuration menus and return to the System OK screen.

At this point, the current temperature of the thermostat should be shown on the thermostat screen.

Part 2: Configuring the Remote Temperature Sensor

In Lab 7-4, "Addressing Dimmers," the remote temperature sensor was connected to terminal Z3. Figure 7-5-2 shows a remote temperature sensor.

Figure 7-5-2 Remote Temperature Sensor

In order for the Omni to be able to understand the information being reported back to this zone, the zone must be correctly configured.

1. To access the installer setup mode, first press the 9 key.

2. Enter the installer code 1, 1, 1, 1.

   This will access the setup menu.

3. Press the # key to access the Installer setup menu.

4. Press the 2 key to access the Zone setup menu.

5. Use the down arrow key to scroll down to Zone 3.

To change a zone type:

1. Press the # key.

2. Use the arrow keys to scroll through the list of zone types.

3. Scroll down until the screen shows #81, OUTDOOR TMP.

4. Press the # key to select this new type.

5. Press the * (asterisk) key repeatedly to exit the Installer configuration menus and return to the System OK screen.

Observe the screen on the thermostat. The current thermostat temperature will be displayed for a couple of seconds, and then the temperature reported from the outdoor temperature sensor will display.

Part 3: Displaying Zone Status

The HAI communicating thermostat and outdoor temperature sensor can be observed from the HAI console.

To observe the status of the thermostat:

1. Press the 6 key (STATUS).

2. Press the 5 key to indicate temperature.

3. Press the 1 key to indicate temperature zone 1.

    When thermostats are connected to the console, they are referred to as *temperature zones*. They are different from *input zones*. Each thermostat connected to the system can be set up in a different temperature zone. By default, the first thermostat will be indicated as zone 1.

4. Press the # key to show the status of thermostat 1.

5. Press the * (asterisk) key repeatedly to exit the menus and return to the System OK screen.

Part 4: Commanding the Thermostat

The thermostat can also be set and/or programmed from the console.

To change thermostat settings from the HAI console:

1. Press the 5 key to enter the temperature settings menu.

2. Press the 1 key followed by the # key to input TEMPERATURE ZONE 1.

On this menu, there are six options:

▶   1: MODE

▶   2: HEAT

▶   3: COOL

▶   4: FAN

▶   5: HOLD

▶   #: STAT

The MODE options change the HVAC system from OFF to HEAT to COOL. This setting indicates the desired method of temperature control, and is used prior to commanding set points with the HEAT and COOL options.

The HEAT and COOL options are used to tell the thermostat what temperature to adjust to, depending on what mode the thermostat is currently in.

Explanation:

If the thermostat is set to HEAT, the thermostat will use the heater to achieve the desired temperature set in the HEAT option. If the thermostat is set to COOL, the thermostat will use the cooler to achieve the desired temperature set in the COOL option.

Example:

In the summer, the thermostat would be set to COOL, and the COOL temperature might be set to 80 degrees. In the winter, the thermostat would be set to HEAT, and the HEAT temperature might be set to 70 degrees. In the fall, when the temperature is just right, the thermostat would be set to OFF.

The FAN control simply uses the fan in the HVAC system to circulate air without heating or cooling it.

The HOLD setting will hold the temperature at the current desired temperature setting, effectively overriding any heating and cooling temperature setpoints or programs.

The # key will simply show the current status of the thermostat.

Part 5: Creating Temperature Set Points

As you work through these scenarios, refer to the previous Parts when making settings. Also, observe the readout on the thermostat's display.

Scenario 1: Winter

Imagine that it is winter and the temperature outside is 30 degrees.

1. Set the thermostat to heat the home to 70 degrees.

Beginning from the System OK screen, write down the steps taken to achieve this setting.

_____

_____

_____

_____

_____

_____

The thermostat will now show that it is on remote, which means that control settings are being sent by the console. The thermostat will also show that it is in heat mode. The thermostat will briefly show the desired temperature that was set: 70 degrees.

Scenario 2: Summer

Imagine that it is summer and the temperature outside is 100 degrees.

1. Set the thermostat to cool the home to 80 degrees.

Beginning from the System OK screen, write down the steps taken to achieve this setting.

_____

_____

_____

_____

_____

_____

2. Turn the thermostat off.

Beginning from the System OK screen, write down the steps taken to achieve this setting.

_____

_____

_____

_____

_____

_____

3. Turn the thermostat back on.

Note: When the thermostat is set to off, remote commands will not be accepted from the console.

Lab 7-5

Motion Detectors

INTRODUCTION

Infrared motion sensors are used to detect objects that generate heat and infrared radiation. Humans give off infrared radiation at a wavelength of between 9 and 10 micrometers because the approximate skin temperature is about 93 degrees F. A typical infrared detector will be sensitive to a range of about 8 to 12 micrometers to cover this range of radiation.

Within the sensor, the infrared radiation emitted by an object bumps electrons off a substrate that are detected by the electronic components of the sensor and amplified into a signal.

The sensor has a difficult time detecting a heat source that is not in motion because the electronics within the sensor look for a rapidly changing amount of heat within the field of its view. This change in energy would indicate a person moving through the field of the sensor's detection.

You must be cautious when installing infrared motion detectors because infrared energy does not pass through glass very well. If you were to install an infrared motion detector to detect motion through a window, such as someone passing by, it would probably not be able to detect a rapid change in heat dissipation. This same sensor can detect a change if a person enters a home through a window.

WHAT YOU WILL DO

▶ Connect the motion detector to the HAI controller board.

▶ Configure the zone type of the motion detector.

WHAT YOU WILL NEED

▶ Motion detector

▶ Category 5e cable

▶ Cable stripper

▶ Snips

▶ Screwdrivers

OPTIONAL ITEMS

▶ URL: http://www.ademco.com/asc/products/
motion_sensors/998.htm

▶ URL: http://www.homeauto.com

Figure 7-6-1 Motion sensor

Setting Up

The first time the motion sensor is installed (see Figure 7-6-1), the knockout holes located around the housing must be opened. Make sure to open one for the Category 5e wire at the top and as many as needed to mount the sensor to the wall. Refer to the instructions included with the sensor when removing the knockouts. The unit can be corner-mounted or flush-mounted, depending on which knockouts are used.

Part 1: Connecting the Sensor

1. Cut a piece of Category 5e cable that is long enough to reach from the controller board to the inside of the sensor with an extra 10 inches (five for slack and five for termination).

2. Remove about one inch of jacket from one end of the cable (this will be the sensor connections) and two inches of jacket from the other (this will be the controller board side).

3. Strip one-quarter inch of each conductor on each end of the cable.

4. Twist the wire pairs back together after they are stripped. This will create four conductors.

5. Connect the brown pair to the plus (+) terminal inside the motion detector.

6. Connect the green pair to the minus (-) terminal.

7. Connect the blue pair to the third terminal and the orange pair to the last terminal. The last two connections provide the connections to the zone on the controller board.

Part 2: Connecting the Sensor to the Controller Board

1. At the other end of the cable, connect the brown pair to the first terminal of the 12V pair on the controller board.

2. Connect the green pair to the second terminal of the 12V pair. These connections will provide the sensor with voltage.

3. Connect the blue pair to Z1 and the orange pair to the terminal next to Z1, marked COM. This will complete the circuit. When the motion sensor detects motion, the circuit is opened. The controller board detects that the circuit has been opened and triggers an alert. The type of alert that is triggered must be programmed into the controller board. A sounder could be activated, a phone call could be made, or an e-mail could be sent to someone upon the event.

Part 3: Setting Up the Zone

In order for the controller board to know what type of device has been connected to it, the zone that the device is connected to must be programmed into the controller.

1. To access the Installer setup mode, first press the 9 key.

2. Enter the installer code 1, 1, 1, 1.

   This will access the setup menu.

3. Press the # key to access the Installer setup menu.

4. Press the 2 key to access the Zone setup menu.

5. Use the down arrow key to scroll down to zone 1.

To change a zone type:

1. Press the # key to indicate change.

2. Use the arrow keys to scroll through the list of zone types.

3. Scroll until the screen shows PERIMETER.

4. Press the # key to select this new type.

5. Press the * (asterisk) key repeatedly to exit the installer configuration menus and return to the System OK screen.

The controller board now knows that the device connected to zone 1 is a perimeter type device.

Part 4: Programming the Controller Board

Now that the sensor is installed and the controller is configured to use the sensor correctly, a program must be set up. For this lab, a simple program will be configured to turn on all the lab wall lights whenever the sensor is tripped while the system is armed.

1. To access the installer setup mode, first press the 9 key.

2. Enter the installer code 1, 1, 1, 1.

   This will access the setup menu.

3. Press the 3 key to access the Program setup menu.

4. Press the 1 key to add a new program.

5. Press the 2 key to indicate that a command will be entered.

6. Press the 4 key to indicate that this will be a program that will affect all X-10 controls.

7. Press the 1 key to indicate that all X-10 devices will be turned on when the alert is tripped.

8. Press the 3 key to access the conditions menu.

9. Press the 3 key to indicate that a zone condition will execute this program.

10. Press the 1 key to indicate which zone will execute this program.

11. Press the # key to input the zone into the program.

12. Press the 1 key to indicate that the zone must be NOT READY to execute the command in the program.

13. Press the # key to display the program.

14. Press the # key to have the controller store the program.

Now that the program has been set up, ensure that the controller has stored the program:

1. Press the 2 key to select SHOW. This will display a screen that allows you to see all the programs stored in the controller.

2. Press the 4 key to indicate that you want to see programs associated with the ALL function.

   If the controller displays ***No Programs***, back out of the menus to the System OK screen and repeat the process to enter the program.

3. Scroll down to see the program that was entered.

   The screen will show the time and date and &Z1N on one line and ALL ON on the next line.

4. After confirming the program entry, back out of the menu system to the System OK screen.

Part 5: Testing the Sensor

Now that the program has been entered, it is time to test it to make sure it will work.

The first thing to do is to clear the area in front of the sensor of any people. Next, the system must be armed. Because this program does not contain any conditions about when it will activate, it will be activated any time the system is armed.

1. To arm the system, press the AWAY key.

2. Press 1, 1, 1, 1 to enter the master code to arm the system.

    The system will beep repeatedly until it is armed.

3. While the system is arming, go to a location in the room either behind the sensor or to the side of it as far away as possible.

4. After the system stops beeping, walk in front of the wall toward the sensor.

    After the sensor has detected motion in front of it, the alarm in the console will go off, the lights on the lab wall will turn on, and the message BURGLARY! ZN1 Perimeter tripped will be displayed on the console.

5. The system must be turned off to silence the alarms. Press OFF.

6. Enter the master code of 1, 1, 1, 1 to disarm the system.

    The system will now indicate SYSTEM OK.

Connectors

INTRODUCTION

The purpose of this lab is to identify, examine, and determine the characteristics of different types of connectors used during cabling installations. It is important to determine the type of connectors that are associated with twisted-pair and coaxial cables in order to install, test, and troubleshoot cable installation jobs.

In order to ensure quality installations and facilitate using the correct connector, it is critical that the installer be aware of the connectors that will be required to terminate cables. This lab will help you master the techniques for knowing which connector to use.

WHAT YOU WILL DO

▶ Identify common cabling connectors.

▶ Touch and examine the characteristics of each connector.

WHAT YOU WILL NEED

▶ Type F connectors

▶ 8P8C connectors

▶ BNC connectors

▶ RCA connectors

Activity

Identify and inspect the different types of cable connectors.

Type F Coax Connectors

The type F snap-and-seal connectors are used to terminate coaxial cable (see Figure 8-1-1). Once terminated with the CATV F crimper, you are ready to connect it to the receptacle. As you connect the completed cable to its receptacle, you will notice some resistance when twisting. The O ring inside will create a watertight seal. The RG6 are snap-and-seal F connectors that we will be using to terminate coaxial cable. The tool that is used to crimp the connector is the SealTite 59/6 Waterproof CATV F Crimp Tool.

8P8C Connectors

The 8P8C connectors provide an eight-contact for twisted-pair cabling (see Figure 8-1-2). The pins are numbered 1 to 8, and different pin combinations are used with different networking standards according to the EIA/TIA standards,

Figure 8-1-1 Type F Coax Connectors

Figure 8-1-2 8P8C connectors

Figure 8-1-3 BNC connectors (RF connector)

Figure 8-1-4 Male RCA connectors

which describe the characteristics and applications for various grades of UTP cabling. Most cabling installations use Category 5e (CAT 5e) twisted-pair cable. CAT 5e cable consists of four-pair (eight wires). A patch cable (straight-through) will be wired and have the same color of wire (pins 1–8) on both ends.

BNC Connectors (RF Connector)

The BNC (British Naval Connector or Bayonet Nut Connector or Bayonet Neill Concelman) is the standard type of connector used to connect IEEE 802.3 10Base 2 coaxial cable. BNC connectors (see Figure 8-1-3) are used with coaxial cables such as the RG-58 A/U cable. A BNC cable has a male type BCN connector at each end. A rotating ring on the outside of the male connector locks the cable to a female BNC connector. During our labs, we will be connecting a coaxial cable with an F connector to a female BNC connector. We will then twist and lock the BNC connector to the LAN ProNavigator cable tester to check for continuity of the cable.

Male RCA Connector

The RCA (Radio Company of America) connectors are a type of unbalanced interconnection used mostly in the video and audio worlds (see Figure 8-1-4). RCA connectors are available as basic connectors as well as precision versions with very tight production tolerances. RCA connectors are mainly used for carrying video and line-level audio on stereo TVs and VCRs. Male RCA connectors have a center pin surrounded by a flower. Female RCA connectors have a center hole and outer cylinder. During our labs, we will be connecting a coaxial cable to a male RCA adapter and our final connection will be to the video sequencer.

Conclusions

During this lab, you identified the different types of connectors that are necessary to complete a cabling installation job. You examined the characteristics of each and you demonstrated your ability to identify the proper connectors for twisted-pair and coaxial cables.

Exercises

Which cable type is associated with the 8P8C connectors? _____

Which type of connector is used with RG6 cable? _____

Which cable type is associated with the F connector? _____

Which type of connector is used with RG-58 cable? _____

Measuring AC

INTRODUCTION

AC outlets are used all the time, both at home and at work. This lab will explain the functions and the voltages present within an AC outlet.

AC outlets are typically connected to three wires. The green or bare copper wire is the grounding conductor, the white wire is the grounded (neutral) conductor, and the black or red wire is the ungrounded (hot) conductor.

The ungrounded conductor is the conductor that feeds power to an appliance. The grounded conductor is the electricity path back to ground, which completes the circuit. The grounding conductor is an extra safety wire that provides an additional path to ground. Current flows through the grounding conductor in the event of a fault. The grounding conductor is normally connected to an appliance housing.

A multimeter is an electrical testing tool capable of detecting voltage levels, resistance levels, and open or closed circuits. A multimeter is capable of checking alternating current, and direct current voltages. Open or closed circuits are shown in ohms, which is a resistance measurement.

Direct current is a voltage that stays at a certain level flowing in one direction; alternating current is a voltage that changes between positive and negative very rapidly.

Warning: All electrical outlets and wires should be treated as if they are hot, whether they have been turned off or not.

WHAT YOU WILL DO

▶   Identify the parts of an AC outlet.

▶   Understand AC outlets.

▶   Use a multimeter safely.

WHAT YOU WILL NEED

Multimeter

OPTIONAL ITEMS

▶   Multimeter instruction book

▶   Provided outlet drawing

Setting Up

Select a working outlet in one of the Structured Media Centers or find a working outlet in the classroom. Figure 9-1-1 illustrates a working outlet and its parts.

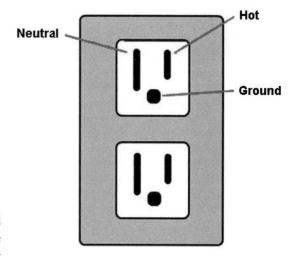

Figure 9-1-1 AC outlet and its parts

Part 1: Familiarizing Yourself with the Meter

1. Set the multimeter to read AC volts. It is very important to set the meter to the correct setting for the application. Some meters can be damaged if they are not set correctly. Figure 9-1-2 shows a multimeter.

   Warning: Do not touch the bare metal parts of the probe!

   The following are some common buttons on multimeters:

   ▸ HOLD. This button is pressed to freeze the present reading on the display.

   ▸ MIN MAX. This button stores minimum, maximum, and average input values.

   ▸ RANGE. This button exits auto ranging and locks on the present range.

   ▸ Hz. This button is pressed to measure the frequency of a voltage or current signal.

Figure 9-1-2 Multimeter

2. Insert the probes of the meter into the slots of an AC outlet. The red probe should be inserted into the shorter slot; the black probe should be inserted into the longer slot.

   What is the reading on the meter?

   _____

Part 2: Testing the Meter

1. Remove the probes from the outlet.

2. Insert the probes, this time with the red probe in the longer slot and the black probe in the shorter slot.

   What is the reading on the meter?

   _____

Part 3: Testing the Ground

1. Leaving the black probe in the shorter slot, remove the red probe and insert it into the round hole in the outlet. This is the grounding connection or safety ground.

   What is the reading on the meter?

   _____

2. Replace the red lead into the longer slot.

3. Remove the black lead and insert it into the round hole, the grounding connection.

   What is the reading on the meter?

   _____

Measuring DC

INTRODUCTION

A modern residence that is wired according to the standard may include cables used for telephony, networking computers, security, and home automation. According to the TIA/EIA 570A standard, this cabling will use a star topology. Wherever they come from, all cables converge at a centrally located panel called the Structured Media Center (SMC). Each TIA/EIA 570A residence will have one or more SMCs. Inside the SMC will be several modules designed to accomplish various home integration tasks. Many of these units need power in order to operate. These modules use commonly available AC adapters. These compact supplies plug into an AC outlet for primary power, and with their internal components they produce a filtered DC output to power modules.

Different modules require different voltages and different polarities of voltage (positive and negative). In addition, different modules require different plug configurations for their DC power input. This means that the installer must be able to discern which supply is appropriate for which application. This is necessary so that modules with different voltage and polarity requirements but similar looking plugs are not inadvertently mixed, possibly destroying the module.

Warning: All electrical outlets and wires should be treated as if they are hot, whether they have been turned off or not.

WHAT YOU WILL DO

▶ Identify the voltage and polarity of an AC adapter using labels and documentation.

▶ Use a multimeter to verify the voltage and polarity of an AC adapter.

WHAT YOU WILL NEED

▶ Multimeter

▶ Several different AC adapters

▶ A 9-volt battery

OPTIONAL ITEMS

Instructions provided with the multimeter

Setting Up

1. Select a working outlet in one of the Structured Media Centers or find a working outlet in the classroom.

2. Set the multimeter to read DC volts. Figure 9-2-1 shows a multimeter.

Figure 9-2-1 Multimeter

It is very important to set the meter to the correct setting for the application. Some meters can be damaged if they are not set correctly. In the case of low-voltage DC measurements, incorrect settings may lead to inaccurate readings.

Part 1: Measuring the Voltage of a Battery

1. Touch the probes of the meter to the 9-volt battery (or any small lantern or flashlight battery on which the terminals are clearly marked). Figure 9-2-2 shows a 9-volt battery.

Figure 9-2-2 9-volt battery

2. Touch the red probe to the terminal marked with a (+) and the black terminal to the terminal marked with a (−).

   Warning: Do not touch the bare metal parts of the probe. It can have a lethal outcome if foreign voltages are present or if you inadvertently do so while measuring AC voltages.

   What is the reading on the meter? _____

Part 2: Reversing the Probes

1. Remove the probes from the battery.

2. Reconnect the probes, this time with the red probe on the negative terminal and the black probe on the positive terminal.

   What is the reading on the meter? _____

If you connected the meter correctly during the steps in Part 1, you should have seen a 9 volts indication when the red probe was on the positive terminal. When you swapped the leads around and placed the black lead on the positive terminal in Part 2, you should have obtained a reading of −9 volts. The meter reads − when the red probe is connected to a negative source. This is one method you can use to verify the polarity of an unmarked AC adapter.

Part 3: Examining the AC Adapter

Obtain a selection (three to five) of power supplies from your instructor (see Figure 9-2-3). In what ways do they differ one from another?

_____

Notice the markings. There is usually some indication of input voltage (typically, 120 VAC) and of the input power line frequency (usually 60 Hz). There is also likely to be an indication of the input power wattage (for example, 15 watts (W), or 15 VA (volt-amps)) as well as a description of the AC adapter's power out, such as 7.5 VDC 1A (7.5 volts, direct current, at 1 ampere maximum current).

For the AC adapters your team has selected, fill in the blanks in the following form:

	Input Voltage	Input Wattage	Output Voltage	Output Current
1	_____	_____	_____	_____
2	_____	_____	_____	_____
3	_____	_____	_____	_____
4	_____	_____	_____	_____
5	_____	_____	_____	_____

Figure 9-2-3 AC adapter

Part 4: Examining the Power Ratings

Input wattage is the amount of power the AC adapter consumes. Output power is the amount of power the unit delivers.

For DC circuits, you can calculate power by multiplying the voltage by the amperage. Thus, an AC adapter marked as 7.5 volts, 1 amp would provide power in the amount of 7.5 watts.

The same AC adapter may be marked as 120 V, 15 watts.

If 15 watts goes in and 7.5 watts comes out, the rest must be absorbed in the process of converting the incoming AC into DC. This means that the adapter is about 50 percent efficient at best. Where do you think the rest of the power goes?

_____

(Hint: After the AC adapter has been plugged in for a while, touch the plastic housing.)

Most of the difference between power-in and power-out is dissipated as heat. This is why it is important to pay attention to the capabilities of an AC adapter. If too much demand is made of it, it could overheat. If this happens, the AC adapter could fail, shutting down a portion of the integrated home.

It is also very important to match the AC adapter to the intended load:

▸ Voltage should match exactly.

▸ Power should be the same or greater.

For instance, a data hub that requires 12 volts and draws 150 mA (150 milliamps, or 0.150 amp) should be fed by an AC adapter that produces 12 VDC. The AC adapter should provide at least 150 milliamps, but one that produces 250 mA or more would do as well. However, an adapter that produces 15 or 9 volts (more or less than specified) should not be used, nor should one that produces less power than the hub requires (say 100 mA).

Part 5: Examining the Connector

Look closely at the connectors on each of the AC adapters. What differences can you describe?

_____

Common differences might be the length of the plug, diameter of the plug, and heaviness (wire gauge) of the connecting wire. Some connectors may not even appear to be a plug at all, but could be a special-purpose connector such as a type F connector (these are used to power certain video distribution modules borrowed from the cable industry). Could differences such as these cause an AC adapter not to fit in every potential module inside a residential Structured Media Center? _____

Size certainly matters in this case, but a factor that is even more important is polarity. How can one tell whether a given AC adapter provides positive or negative current to the center pin or shell of the connecting plug?

_____

Most AC adapters have a symbol such as the following on them, shown in Figure 9-2-4.

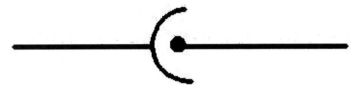

Figure 9-2-4 AC adapter symbol

This symbol indicates that the center pin of the adapter connector is positive, and that the surrounding shell is negative.

For the AC adapters you examined in step 3, note in the following form whether or not the positive terminal is on the center pin or the shell.

	Center Positive?	Center Negative?
1	_____	_____
2	_____	_____
3	_____	_____
4	_____	_____
5	_____	_____

From this, would you say that all AC adapters have a positive center pin? _____

Could you say that most AC adapters have the center pin positive? _____

What could you do to find out the voltage and polarity of an AC adapter that came without a label or with an obscured label?

_____

_____

Part 6: Using Test Equipment to Verify the Connector

Using a multimeter and the same techniques you used in step 1, measure the output voltage of each AC adapter. Touch the red lead to the center pin and the black lead to the shell. Is the resulting voltage positive or negative? Record your results in the following form:

AC Adapter	Voltage Measured	Center Pin Positive?	Center Pin Negative?
1	_____	_____	_____
2	_____	_____	_____
3	_____	_____	_____
4	_____	_____	_____
5	_____	_____	_____

How do these measurements compare with the observations you made in Part 3 and Part 5?

_____

_____

_____

HAI X-10 Controls

INTRODUCTION

X-10 is a power line carrier protocol that allows compatible devices throughout the home to communicate with each other via the existing 110 V wiring in the house. Using X-10, it is possible to control lights and virtually any other electrical device from anywhere in the house with no additional wiring!

X-10 is a communications "language" that allows compatible products to talk to each other via the existing 110 V electrical wiring in the home. No costly rewiring is necessary.

X-10 transmitter devices send a coded low-voltage signal that is superimposed over the 110 VAC current. Any X-10 receiver device plugged into the household 110 V power supply will see this signal. However, the receiver device responds only when it sees a signal that has its address. Up to 256 different addresses are available. If you want more than one device to respond to the same signal, simply set them to the same address.

X-10 devices can be categorized into three distinct groups:

▶ Transmitters (see Figure 10-1-1)

▶ Receivers (see Figure 10-1-2)

▶ Transmitter/receivers (two-way X-10 devices) (see Figure 10-1-3)

WHAT YOU WILL DO

▶ Learn how to use the HAI console to control X-10 lights.

▶ Learn how to use the HAI console to program lighting "scenes."

WHAT YOU WILL NEED

HAI owner's manual

OPTIONAL ITEMS

▶ URL: http://leviton.com/sections/prodinfo/newprod/npleadin.htm

▶ URL: http://www.homeauto.com

▶ URL: http://www.x10.org

▶ URL: http://www.x10.com

▶ URL: http://www.x10ideas.com

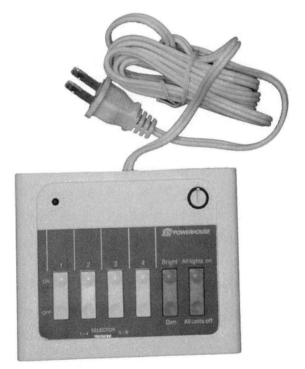

Figure 10-1-1 Transmitter

Figure 10-1-2 Receiver

SETTING UP

Make sure that the HAI console has been set up and that it is functioning properly. Double-check the console settings if necessary or ask your instructor for help if you are having problems.

This lab needs to be performed after the lab that assigns each light an address. If that lab has not been completed yet, complete programming of the lights using that lab before proceeding. The OmniLT will be able to send commands to the individual devices based on the house code number assigned to each switch.

Part 1: Using the All Key

The ALL key is used to turn all lights and X-10 controllers on. It is also used to set scenes, issue a scene on command, and issue a scene off command to switches that support Leviton scene control.

1. Press the ALL key on the HAI console. This is the number 4 key.

2. Press the 1 key on the console. All of the lights that have been set to any X-10 address will turn on.

3. Press the ALL key again.

4. Press the 0 key on the console. All the lights that have been set to an X-10 address on the same house code (Default A) will turn off.

If the lights do not respond to the command, double-check the console settings if necessary or ask your instructor for help if you are having problems.

Part 2: Using the STATUS Key

Note: This key will display the status of X-10 devices that have been set using a common X-10 controller or the HAI console. There are some other dimmer switches that can actually report back to the HAI console the current status of the device when they are turned off locally (manually).

The current status of any controllable two-way X-10 device can be displayed by using the STATUS or 6 key on the console.

1. Turn off all the lights on the lab wall.

2. Press the STATUS key.

3. Press the 1 key, followed by the # key to display the status of device #1.

   What is the status of device #1?

   _____

   _____

   _____

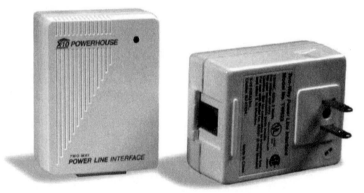

Figure 10-1-3 Transmitter/receivers (two-way X-10 devices)

4. Press the * (asterisk) key. This will return you to the previous controller menu.

5. Press the 2 key, followed by the # key to display the status of device #2.

What is the status of device #2?

_____

6. Turn on light #2 on the lab wall using the physical dimmer (light switch).

What is the status of device #2?

_____

Why is this displayed?

_____

Part 3: Using the CONTROL Key
The CONTROL key is used to control individual lights and appliances.

1. Press the CONTROL key on the console. This is the 1 key.

2. The first named item in the items list will be displayed automatically. The arrow keys can be used to scroll through the list of named items in order to choose a specific item. If the specific number is known, enter the item number followed by the # key to display the item.

   After the unit has been selected, the console will display the following options that can be performed on the device:

   0   will turn the item off.

   1   will turn the item on.

   2   will bring up the dimming options.

   3   will bring up the brightening options.

   4   will show lighting-level options.

   5   will show ramp options for ALC appliances.

   9   will bring up timing commands for the item.

   #   will show the status of the item.

3. Experiment turning each light on, off, dim, and bright. To return to a previous menu, press the *(asterisk) key. The lab wall should have six different lights that can be controlled by using items 1–6 in the list.

4. When you are finished, return to the System OK screen.

Part 4: Programming a Scene
The HAI console is capable of setting lighting scenes. A lighting scene is a group of up to four lights that can be set to different levels that are remembered by the console. Each lighting scene can be completely different, from all lights on to all lights off or anything in-between if the switches are dimmer-capable. If a switch is not dimmer-capable, the console will only be able to turn the device on or off. 256 different scenes can be set and used from the HAI console. Four different scenes can be set to each set of four lights. The first four scenes (1–4) control the first four (1–4) lights. The next four scenes (5–8) control the next four (5–8) lights, and so on.

To set a lighting scene:

1. Press ALL on the console.

2. Press 2 for SCENE.

3. Enter the scene number you want to control (1).

4. Press the # key to accept the choice.

5. Adjust the first four (1–4) lights on the lab wall to varying dimness, or on or off status if they are not dimmer controls.

6. Press the 2 key on the console to program the scene.

Part 5: Testing a Scene

Now that scene #1 has been programmed, test the scene:

1. Press ALL.

2. Press 0 to turn off all the lights.

3. Press ALL.

4. Press 2 to access the scenes menu.

5. Press 1 followed by the # key to access scene #1.

6. Press 1 to turn the scene on.

7. Press ALL to access the ALL X-10 devices menu.

8. Press 0 to turn off all the lights.

Part 6: Programming Different Scenes

1. Program Scene #2 with different lighting levels for each light than the levels used in Scene #1. Use the steps in Part 3 for help, exchanging 1 for 2 when asked for the scene number.

2. Follow the steps in Part 4 to test the scene, exchanging 1 for 2 when asked for the scene number.

3. Switch from one scene to the other by turning on Scene #1 and then turning on Scene #2. There is no need to turn off the lights between changing scenes.

4. Program Scene #3 with yet other lighting levels.

5. Switch among the three scenes to display each one.

6. Turn off all the lights using the console.

HAI Software Installation

INTRODUCTION

PC Access software allows the homeowner to access Home Automation's Omni automation systems over the telephone or over a direct serial connection (RS-232/RS-485) through a serial interface module by using a PC. The system can be accessed both remotely and locally.

Once installed, the following functions can be performed:

▶ Create and maintain customer account files

▶ Download an account file to a customer's system

▶ Upload an account file from a customer's system

▶ Print an account file

▶ Create/view/modify programs

▶ View/modify unit, zone, button, code, area, and message names

▶ View/modify unit, zone, button, code, area, and message voice descriptions

▶ View/modify setup information

▶ View event log

▶ View status of system, zones, units, and temperatures

The customer, installer, or technician can send control, security, button, temperature, and message commands to a customer's system from anywhere.

WHAT YOU WILL DO

▶ Install software.

▶ Configure modem/RS-232 interface.

▶ Set the security stamp.

▶ Set the password.

WHAT YOU WILL NEED

▶ IBM compatible PC or laptop

▶ Serial port

▶ PC Access software CD

▶ HAI serial cable (21A05-2)

OPTIONAL ITEMS

▶ URL: http://www.homeauto.com

▶ PC Access software manual

Section 1: Software Installation and Modem Configuration

Part 1: Getting Started

1. Insert the 1105/1106 Dealer PC Access CD into your CD-ROM drive

2. Click on the Start button and select the Run... command.

3. Type <drive>:setup

    (Example: d:setup for a CD-ROM drive associated with drive D)

4. At the HAI Dealer PC Access Setup screen, enter the following information (see Figure 10-2-1):

    ▸ User Name: USER

    ▸ Company Name: CLI

    ▸ Serial Number: (located on the CD sleeve)

5. At the Choose Destination Location prompt, select the default location by clicking on the Next button.

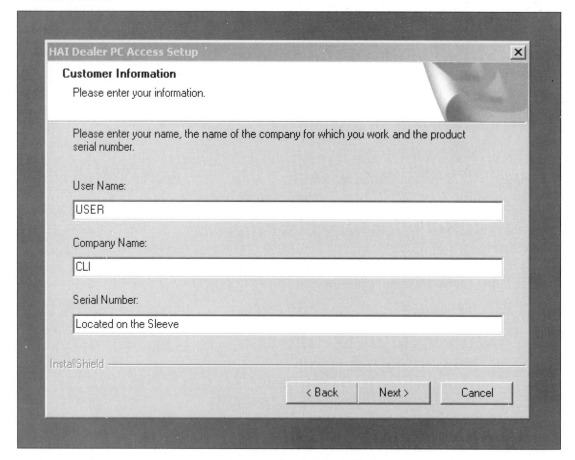

Figure 10-2-1 Installation

Part 2: Installing the Software

1. At the Start Copying Files prompt, verify the Target Directory and User Information. If this matches what you previously entered, select the Next button to begin program installation.

2. At the InstallShield Wizard Complete prompt, click on the Finish button to exit the installation program.

3. Verify that the Dealer PC Access software shortcut has been created on the desktop.

4. Click on Start, Programs, HAI, PC Access; and verify that a program entry has been made for the Dealer PC Access software.

Part 3: Starting Dealer Access Software for the First Time

Note: When running the PC Access program for the first time, the "Install PC Access" dialog box is displayed and prompts you to enter a security stamp. The security stamp can be anything you choose. You will only have to enter a security stamp when installing the program on a computer for the first time. This security stamp is used to encrypt your account files so that only you can access them.

Figure 10-2-2 Security stamp

1. Double-click the Dealer PC Access software shortcut on the desktop.

2. Enter Security Stamp. Use 1234 as the security stamp. Figure 10-2-2 shows a screenshot of the security stamp.

   A welcome screen is displayed, which gives information about the program and prompts you to enter your login password.

3. Enter the default password "PASSWORD". Figure 10-2-3 shows a screenshot of the login password screen.

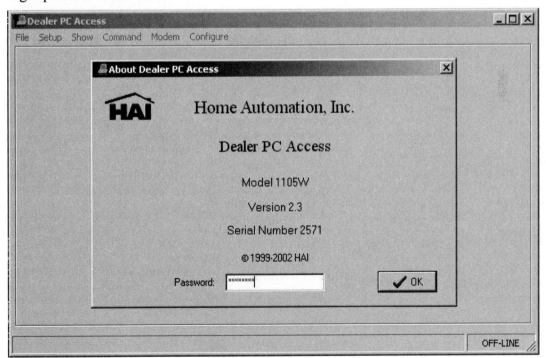

Figure 10-2-3 Login screen

Part 4: Configuring a Modem

The program is initially configured to use a communications device on the COM1 communications port.

The *Configure|Modem* menu can be used to change the communications port, baud rate, and the various modem command strings. The default modem command strings have been found to work well with most Hayes-compatible modems.

When communicating with a controller through a Serial Interface Module, use the *Configure|Modem* menu to change the communications port to the computer's serial port being used by the controller. To select a different baud rate, enter the new rate between 75 and 19200:

1. Click on Configure, Modem. Make sure that Port is set to COM1 (or the Com port you are using for the serial interface) and Baud Rate is set to 9600. Figure 10-2-4 shows a screenshot of the modem configuration.

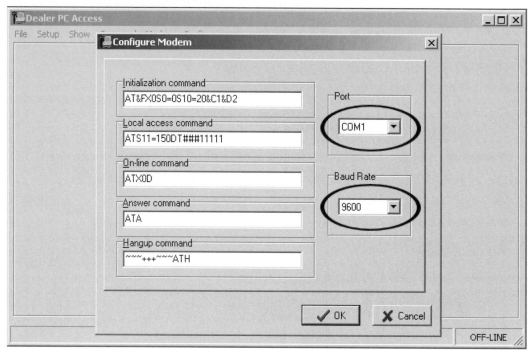

Figure 10-2-4 Modem configuration

2. Click on OK to exit the Modem Configuration screen.

Part 5: Setting Up an Account File

The PC Access software uses a series of data files called "account files." Each account file stores information for a single customer's system. These files are stored on the computer, but the information in them may be downloaded to the customer's system. Similarly, information from the customer's system may be uploaded for storage on the PC. The PC Access software is used to maintain these account files and to send commands to the customer's system.

1. Select File, New.

2. The New Account File prompt will be displayed.

3. Type automation for the filename and click on the Save button. Figure 10-2-5 shows a screenshot of setting up an account file.

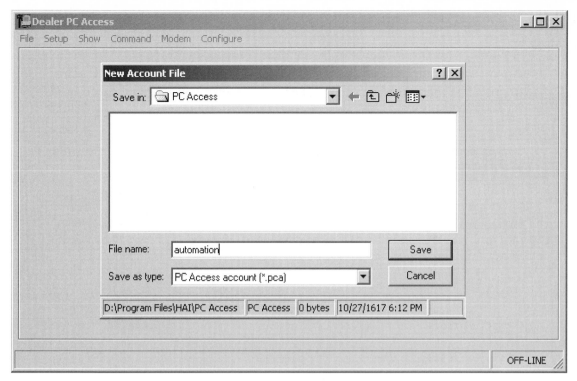

Figure 10-2-5 Setting up an account file

4. At the New Account Model prompt, select the radio button for Omni LT and then click on OK to exit. Figure 10-2-6 is a screenshot of the New Account Model prompt.

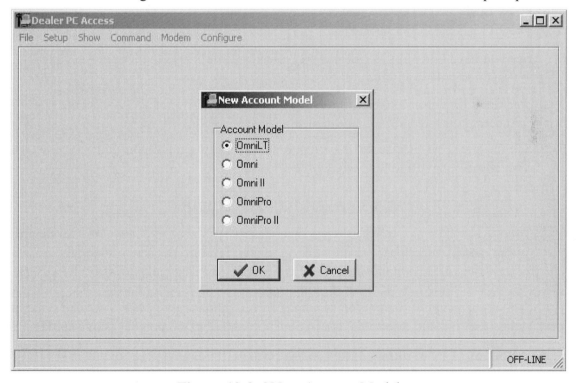

Figure 10-2-6 New Account Model prompt

5. Enter Name, Address, Phone Number, and any remarks necessary for this account. Figure 10-2-7 shows the information screen account.

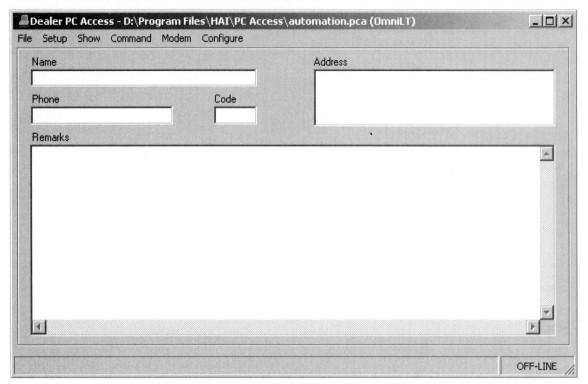

Figure 10-2-7 Account information

6. After you have entered this information, click on File, Save to ensure that your data has been saved to your configuration file.

   Note: Each time you make configuration changes, remember to save them. The Dealer PC Access program does not automatically save file data.

7. Click on File, Exit. If changes have been made to the account configuration file and not saved, click on Save and the Save before Closing prompt.

Section 2: Account Setup

In this section you will create several types of codes. The Manager codes can arm/disarm the security system, access functions that are code-protected in High Security mode, and access the system from an outside telephone line. User codes can be used only to arm and disarm the security system in assigned areas when the time assigned to that code is valid. The Duress code is used by the homeowner to disarm the system if forced against his or her will. The system will perform a silent dial-out and inform the alarm company that the homeowner is in trouble.

Part 1: Opening a Previously Created Account File

1. Double-click the Dealer Access software shortcut on the desktop.

   A welcome screen is displayed, which gives information about the program. You are prompted to enter your login password.

2. Enter the default password "PASSWORD."

3. Select File, Open.

4. In the Open Account File prompt, select the file Automation created earlier; then click on the Open button.

Part 2: Creating a Manager Code

There are 99 user codes that you may assign to users of the system. All Omni LT codes are four digits long. A code can be any number from 0001 to 9999. Each user should be assigned a security code with an authority level, areas that can be accessed (if area arming is used), and times and days in which the code will be valid.

The levels of authority that you can assign to a user code are Master, Manager, User, and Duress.

Master Code

The Master code allows complete access to the entire system. Only the owner(s) or the person(s) who will govern the system should have and use the Master code. A Master code is allowed access to all areas, all the time.

User Code 1 is always set to a Master code of 1111 by default.

Manager Code

The Manager code can arm/disarm the security system in assigned areas during assigned times. The Manager code can access functions that are code-protected in High Security mode. Managers may also access the system from an outside telephone line.

User Code

User codes can be used only to arm and disarm the security system in assigned areas when the time assigned to those codes is valid.

Duress Code

If the homeowner is forced to disarm the system against his or her will by an intruder, they would use the Duress code instead of the normal code. The system will disarm normally. No sirens will sound and no lights will flash, but the system will perform a silent dial-out and say that this is a silent alarm.

Follow these steps to create a manager code:

1. Select Setup, Codes.

   Codes 1 to 99 plus the Duress code are displayed. The Master code is configured as Code 1. The default value of Code 1 is 1111.

   Note: In the lab environment, do not change this code.

2. In the Setup Codes window, select Code 2 in the Code/User Section.

3. In the Setup Codes window, set Code 2 to 1234.

4. In the Setup Codes window, set Code 2 "Authority" to the Manager radio button.

5. In the Setup Codes window, set Code 2 "When on" to read 12:00 a.m.

6. If not already selected, place check marks next to all days (Monday through Sunday).

7. In the Setup Codes window, set Code 2 "When off" to read 12:00 a.m.

8. If any days are selected, uncheck marks next to all days (Monday through Sunday).

9. Click on OK to finish the account setup process.

10. To verify Code 2 settings, select Setup, Codes and select Code 2.

Check settings created for Code 2.

11. Click on OK to exit.

12. Select File, Save to save changes to the Automation file.

Part 3: Creating a User Code

Note: In the lab environment, do not change this code.

1. Select Setup, Codes (see Figure 10-2-8).

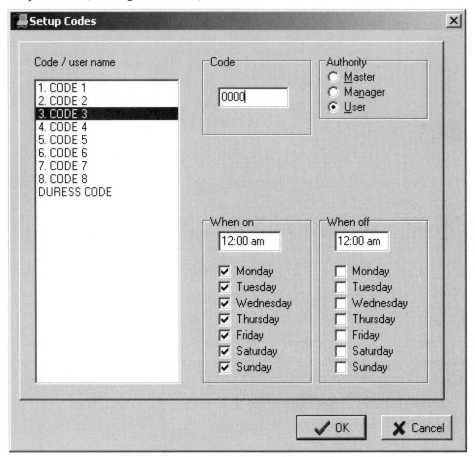

Figure 10-2-8 User code screenshot

2. In the Setup Codes window, select Code 3 in the Code/user section.

3. In the Setup Codes window, set Code 2 to 2222.

4. In the Setup Codes window, set Code 2 "Authority" to the User radio button.

5. In the Setup Codes window, set Code 2 "When on" to read 9:00 a.m.

6. Place check marks next to Monday, Wednesday, and Friday only.

7. In the Setup Codes window, set Code 2 "When off" to read 1:00 p.m.

8. Place check marks next to Monday, Wednesday, and Friday only.

9. Click on OK to finish the account setup process.

Note: This creates an account that is valid only for the days Monday, Wednesday, and Friday; and will work only between the hours of 9:00 a.m. and 1:00 p.m.

1. To verify Code 3 settings, select Setup, Codes and select Code 3.

   Check the settings to make sure they are correct.

2. Click on OK to exit.

3. Select File, Save to save changes to the Automation file.

Part 4: Creating a Duress Code
Note: In the lab environment, do not change this code.

1. Select Setup, Codes (see Figure 10-2-9).

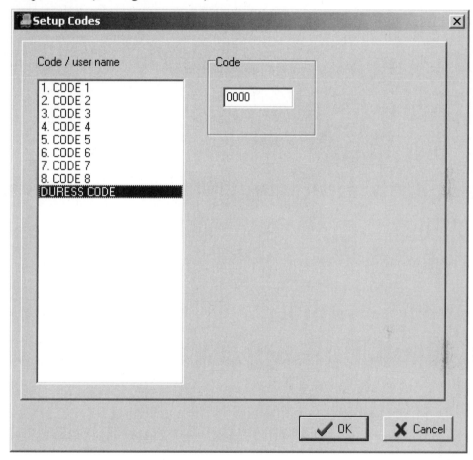

Figure 10-2-9 Duress Code Screenshot

2. In the Setup Codes window, select Duress Code in the Code/user name section (Duress is below Code 99).

3. In the Setup Codes window, set Code 2 to 1911.

4. Click on OK to finish the account setup process.

5. To verify Duress Code settings, select Setup, Codes and select Duress.

6. Check settings created in steps 1 to 4.

7. Click on OK to exit.

8. Select File, Save to save changes to the Automation file.

Part 5: Setting Up Phone Data

1. Select Setup, Dial (see Figure 10-2-10).

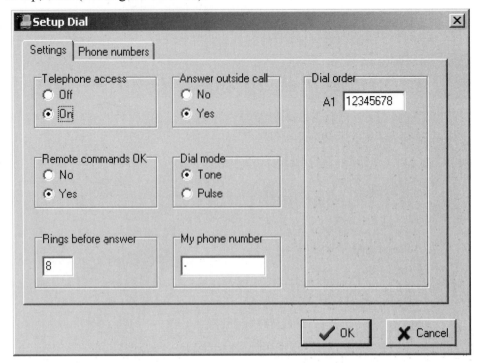

Figure 10-2-10 Phone data screenshot

2. In the Setup Dial dialog box, select the Settings tab and set the following options:

   ▶ Telephone access: On

   ▶ Answer outside call: Yes

   ▶ Remote commands OK: Yes

   ▶ Dial mode: Tone

   ▶ My phone number: 555-6001 (or the number provided to you by your instructor)

   ▶ Dial order: 12345678

3. In the Setup Dial dialog box, select the Phone numbers tab.

4. Phone Number 1 should be set to the security provider's phone number. For lab purposes, enter 555-6003.

5. "When on" to read 12:00 a.m. (If not already selected, place check marks next to all days [Monday through Sunday].)

6. "When off" to read 12:00 a.m. (If any days are selected, uncheck marks next to all days [Monday through Sunday].)

7. Click on OK to finish the account setup process.

8. To verify Code Duress settings, select Setup, Dial. Select the Settings button and Phone numbers button.

9. Check the settings that were created.

10. Click on OK to exit.

11. Select File, Save to save changes to the Automation file.

Part 6: Setting Up Arming

1. Select Setup, Arming (see Figure 10-2-11).

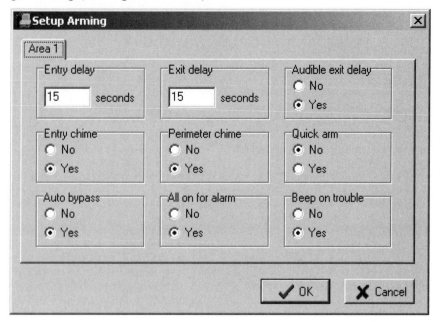

Figure 10-2-11 Arming Screenshot

2. In the Setup Arming dialog box, select the Area 1 tab.

3. Check options and make sure that the following defaults are selected:

   ▶ Entry delay: 15 seconds

   ▶ Entry chime: Yes

   ▶ Auto bypass: Yes

   ▶ Exit delay: 15 seconds

   ▶ Perimeter chime: Yes

   ▶ All on for alarm: Yes

   ▶ Audible exit delay: Yes

   ▶ Quick arm: No

   ▶ Beep on trouble: Yes

4. Click on OK to finish the account setup process.

5. Select File, Save to save changes to the Automation file.

Understanding Blueprints

INTRODUCTION

The purpose of this lab is to introduce you to the different types of blueprints that you will use in the field. These will normally be architectural and electrical prints. Floor plans are used for the basic layout of systems. With the capability to understand these blueprints, you will know what types of outlets to install and where to install them. Blueprints use standard symbols to identify outlet locations and types of outlets.

WHAT YOU WILL DO

▸ View floor plans and system distribution plans.

▸ Identify symbols associated with home integration.

Part 1: Examining a Simple Floor Plan

The drawing shown in Figure 11-1-1 is a simple floor plan for a three-bedroom, single-family dwelling. In most cases, it is obvious what the different components are: doors, windows, and so on. If this were a blueprint, the living details such as the furniture would not be shown. Familiarize yourself with this type of floor plan.

Figure 11-1-1 Simple floor plan for a three-bedroom, single-family dwelling

Part 2: Examining a Telephone and Data Subsystem Drawing

The drawing shown in Figure 11-1-2 illustrates a telephone and data subsystem for a home. Familiarize yourself with this type of drawing.

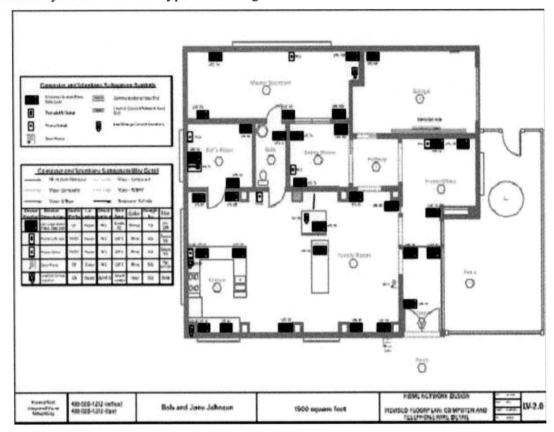

Figure 11-1-2 Telephone and data subsystem

Part 3: Understanding Electrical Symbols

Figure 11-1-3 (Parts 1–3) illustrates various symbols found on blueprints. Some of these symbols are for electricians; some are for low-voltage or home technology integration installers.

1. Place the letter E next to symbols that are for electricians.

2. Place the letter L next to low-voltage or home integrator symbols.

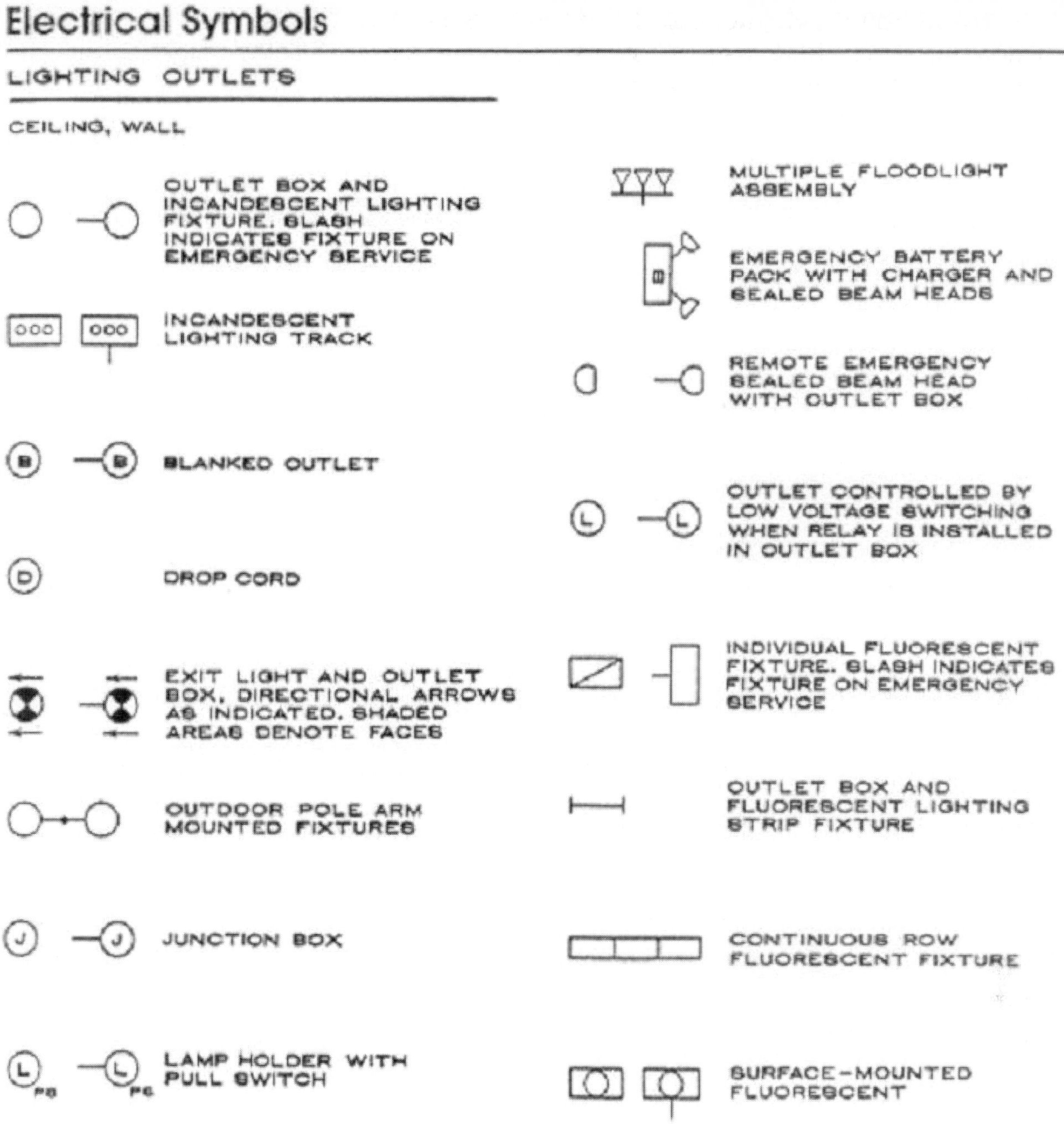

Electrical Symbols

LIGHTING OUTLETS

CEILING, WALL

OUTLET BOX AND INCANDESCENT LIGHTING FIXTURE. SLASH INDICATES FIXTURE ON EMERGENCY SERVICE

INCANDESCENT LIGHTING TRACK

BLANKED OUTLET

DROP CORD

EXIT LIGHT AND OUTLET BOX, DIRECTIONAL ARROWS AS INDICATED. SHADED AREAS DENOTE FACES

OUTDOOR POLE ARM MOUNTED FIXTURES

JUNCTION BOX

LAMP HOLDER WITH PULL SWITCH

MULTIPLE FLOODLIGHT ASSEMBLY

EMERGENCY BATTERY PACK WITH CHARGER AND SEALED BEAM HEADS

REMOTE EMERGENCY SEALED BEAM HEAD WITH OUTLET BOX

OUTLET CONTROLLED BY LOW VOLTAGE SWITCHING WHEN RELAY IS INSTALLED IN OUTLET BOX

INDIVIDUAL FLUORESCENT FIXTURE. SLASH INDICATES FIXTURE ON EMERGENCY SERVICE

OUTLET BOX AND FLUORESCENT LIGHTING STRIP FIXTURE

CONTINUOUS ROW FLUORESCENT FIXTURE

SURFACE-MOUNTED FLUORESCENT

Figure 11-1-3 Electrical symbols (Part 1)

Electrical Symbols

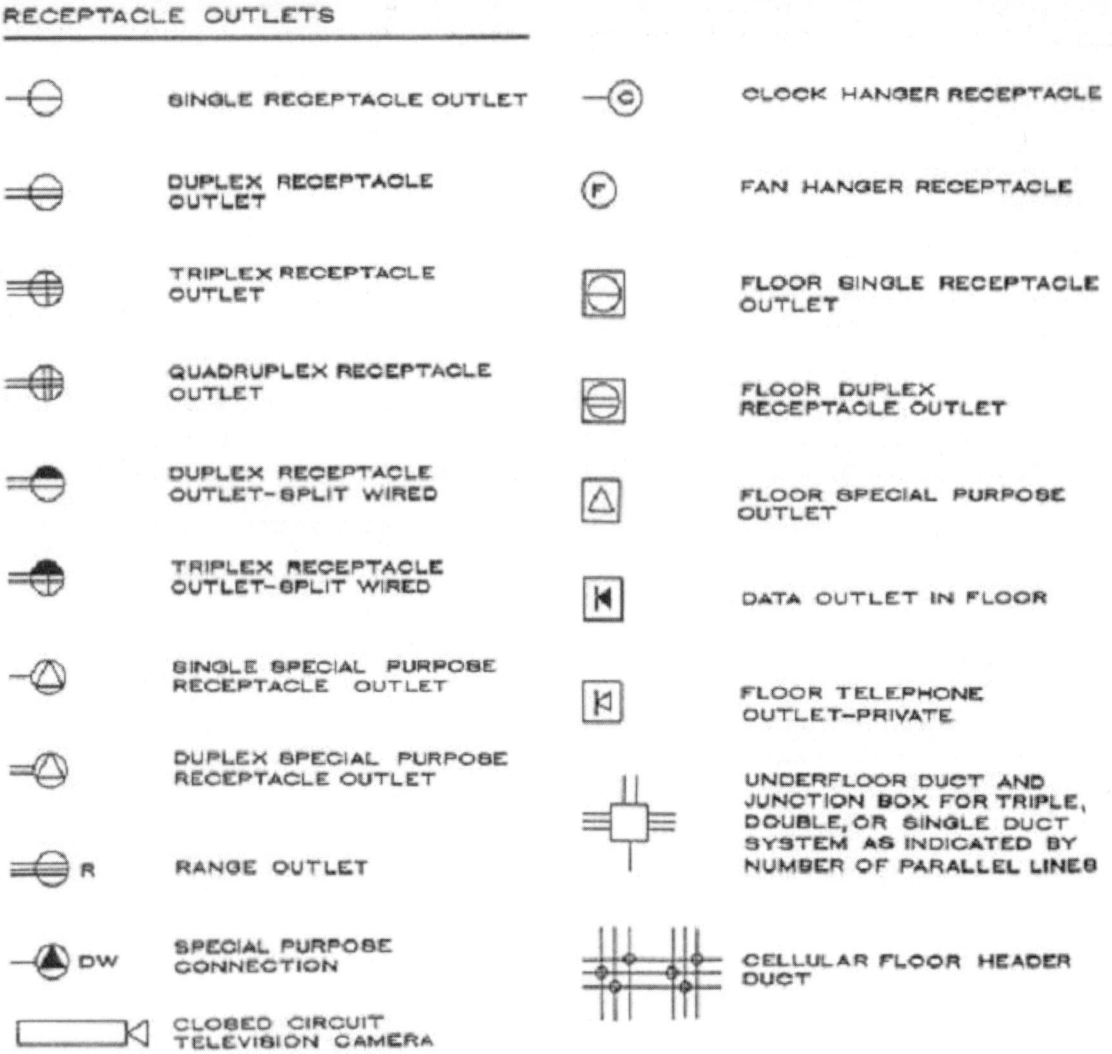

Symbol	Description	Symbol	Description
	SINGLE RECEPTACLE OUTLET		CLOCK HANGER RECEPTACLE
	DUPLEX RECEPTACLE OUTLET	F	FAN HANGER RECEPTACLE
	TRIPLEX RECEPTACLE OUTLET		FLOOR SINGLE RECEPTACLE OUTLET
	QUADRUPLEX RECEPTACLE OUTLET		FLOOR DUPLEX RECEPTACLE OUTLET
	DUPLEX RECEPTACLE OUTLET-SPLIT WIRED		FLOOR SPECIAL PURPOSE OUTLET
	TRIPLEX RECEPTACLE OUTLET-SPLIT WIRED		DATA OUTLET IN FLOOR
	SINGLE SPECIAL PURPOSE RECEPTACLE OUTLET		FLOOR TELEPHONE OUTLET-PRIVATE
	DUPLEX SPECIAL PURPOSE RECEPTACLE OUTLET		UNDERFLOOR DUCT AND JUNCTION BOX FOR TRIPLE, DOUBLE, OR SINGLE DUCT SYSTEM AS INDICATED BY NUMBER OF PARALLEL LINES
R	RANGE OUTLET		
DW	SPECIAL PURPOSE CONNECTION		CELLULAR FLOOR HEADER DUCT
	CLOSED CIRCUIT TELEVISION CAMERA		

Figure 11-1-3 Electrical symbols (Part 2)

Electrical Symbols

SWITCH OUTLETS

S	SINGLE POLE SWITCH	S_{WF}	WEATHERPROOF FUSED SWITCH
S_2	DOUBLE POLE SWITCH		
S_3	THREE-WAY SWITCH	S_L	SWITCH FOR LOW VOLTAGE SWITCHING SYSTEM
S_4	FOUR-WAY SWITCH	S_{LM}	MASTER SWITCH FOR LOW VOLTAGE SWITCHING SYSTEM
S_D	AUTOMATIC DOOR SWITCH	S_T	TIME SWITCH
S_K	KEY OPERATED SWITCH		
S_P	SWITCH AND PILOT LAMP	Ⓢ	CEILING PULL SWITCH
S_{CB}	CIRCUIT BREAKER		SWITCH AND SINGLE RECEPTACLE
S_{WCB}	WEATHERPROOF CIRCUIT BREAKER		SWITCH AND DOUBLE RECEPTACLE
S_{DM}	DIMMER		
S_{RC}	REMOTE CONTROL SWITCH	A,B,C ETC.	
S_{WP}	WEATHERPROOF SWITCH	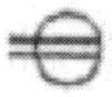 A,B,C ETC.	SPECIAL OUTLETS
S_F	FUSED SWITCH	S A,B,C ETC.	

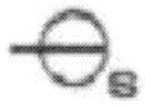

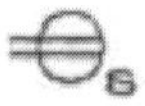

Figure 11-1-3 Electrical symbols (Part 3)

Configuring the Structured Media Center

INTRODUCTION

The SMC is designed to provide a central distribution point for all low-voltage cables. The SMC provides connectivity for the following services: telephones, security cameras, coaxial cable and satellite video distribution, speakers, and volume controls. It also allows for networking, Internet, and whole-house audio sound. Figure 12-1-1 shows the SMC.

The SMC is designed to be flush-mounted between standard 16-inch walls; however, it may also be surface-mounted. It is important that you take into consideration and allow for sufficient space above and below the unit for cable routing and power modules. You may have to consider more than one SMC if the home is large and has many services. A large SMC enclosure can comfortably accommodate up to 12 different distribution modules, depending on the selection and density. It also complies with all TIA/EIA 568-A and TIA/EIA 570-A standards.

The SMC has the following features:

▶ Multiple knock-outs on the top, bottom, sides, and back facilitate cable entry and routing from every desired room.

▶ It provides multiple functionality for wired homes due to its large capacity.

▶ It is made of one piece of sturdy steel construction.

▶ It has hinged covers and will mount on the right or left side for easy access with an optional lock and key for added security.

▶ It can hold a broad range of modules and allows for future distribution modules.

The following recommendations should be taken into consideration when you are planning for the placement of your modules in the SMC: It is important that you carefully plan and design for appropriate placement of your distribution modules. You will need to consider and allow for cable routing placement. Certain modules require that you mount them to the SMC either vertically or horizontally, thus allowing gravity to hold the module securely in place. Other modules allow you to rotate them 90 degrees, which enables neater cable routing. Most modules use push lock pins. You will need to position the module; then align the push lock pins on the distribution panel with the appropriate mounting holes on the enclosure. Seat the panel in place, resting it flat against the back of the

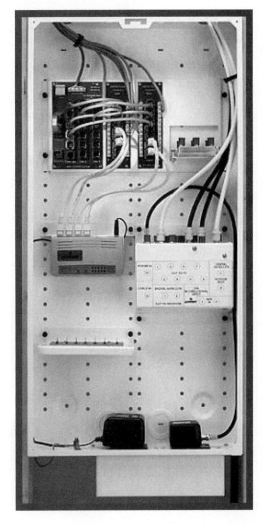

Figure 12-1-1 SMC

panel, and then press the push lock pins inward with your thumb to lock them into place. You will know the push locks are secure if they "click" when you press in. You will find documentation of the characteristics of each module as follows.

▶ Advance Telephone and Video Panel (see Figure 12-1-2). This panel combines multiple modules into one and includes the voice and data module, the telephone distribution module, and the six-way coaxial splitter. The telephone distribution module and the voice and data modules are mounted to the panel with screws. The coaxial splitter is mounted with two nuts and bolts to the panel. After these modules are mounted to the panel, you are ready to affix the telephone and video panel to the SMC. This module must be mounted horizontally. Seat the panel in place, resting flat against the back of the panel, and then press the push lock pins inward with your thumb to lock them into place.

Figure 12-1-2 Advance telephone and video panel

▶ Power Distribution Module (see Figure 12-1-3). This module, which can be positioned vertically or horizontally, has two push pins to secure it into place. However, it will be easier to complete cable runs if it is placed horizontally.

Figure 12-1-3 Power distribution module

▶ Speaker Connection Module (see Figure 12-1-4). This module, which can be positioned vertically or horizontally, has two push pins to secure it into place.

Figure 12-1-4 Speaker connection module

▶ Video Sequencer (see Figure 12-1-5). This module can be positioned vertically or horizontally. Insert the push pin bases into the module's bracket holes, align, and use the push pins to affix the metal bracket onto the enclosure. You should position this module in whichever alignment best meets your needs for the cable runs and the other devices within the enclosure it needs to connect with.

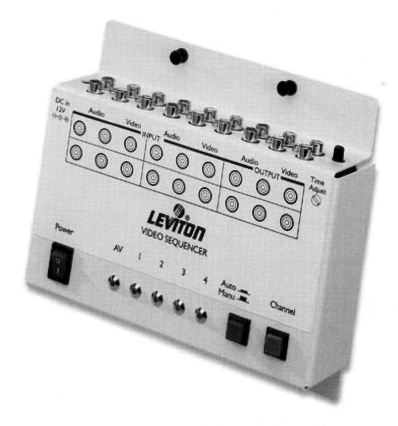

Figure 12-1-5 Video sequencer

Configuring the Structured Media Center

▶ 10Base-T Network Hub (see Figure 12-1-6). This module can be positioned vertically or horizontally. To install the module into the SMC, simply align the mounting pins with the grid holes in the back of the SMC. Secure it by pushing the two push pins into place.

Figure 12-1-6 10Base-T network hub

▶ Internet Gateway (see Figure 12-1-7). This module can be positioned vertically or horizontally. The gateway has two push pins and two _____ for placement in the SMC.

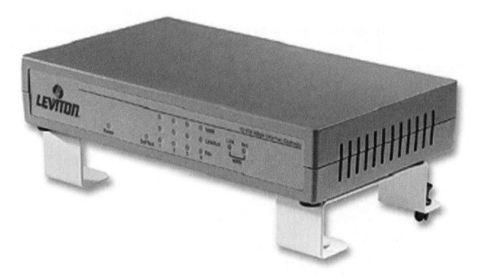

Figure 12-1-7 Internet gateway

▶ Enhanced RT Distribution Module. This module can be positioned vertically or horizontally. You should position this module in whichever alignment best meets your needs for the cable runs and the other devices within the enclosure it needs to connect with.

▶ Enhanced Bi-Directional RF Amplifier. This module can be positioned vertically or horizontally. It has four push pins to secure it into place.

Figure 12-1-8 shows two Enhanced RT Distribution Modules on the top row and an Enhanced Bi-Directional RF Amplifier on the bottom row.

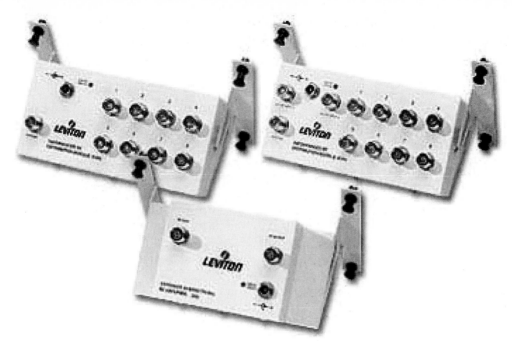

Figure 12-1-8 Enhanced RT distribution modules (top row) and enhanced bi-directional RF amplifier (bottom row)

▶ Decora Media System Hub (see Figure 12-1-9). This module, which can be positioned vertically or horizontally, has four push pins to secure it into place. However, horizontal installations allow access to this module from the top or bottom.

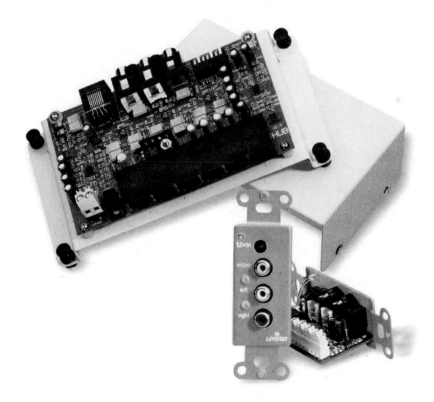

Figure 12-1-9 Decora media system hub

▶ Digital Volume Control Interface (see Figure 12-1-10). This module, which can be positioned vertically or horizontally, has four push pins to secure it into place.

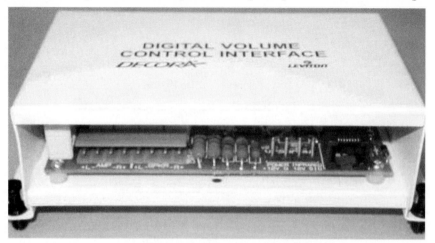

Figure 12-1-10 Digital volume control interface

▶ Video Modulator (see Figure 12-1-11). This module must be mounted horizontally; otherwise, it will slide out of the unit. It uses two push pins to secure it to the SMC.

Figure 12-1-11 Video modulator

▶ AC Power Module (see Figure 12-1-12). This panel secures to the SMC with four screws. Two cable knockouts on the bottom of the panel allow for power source connections. It is recommended that a qualified electrician complete the wiring and grounding process.

Figure 12-1-12 AC power module

WHAT YOU WILL DO

▸ Read an explanation of how you would install the SMC in an actual setting.

▸ Identify the modules and their unique characteristics.

▸ Demonstrate removing and affixing a module to the SMC.

WHAT YOU WILL NEED

Leviton Series 280 Structured Media Center with the modules affixed to the panel, as shown in Figure 12-1-13.

Part 1: Installing the SMC

Part 1 is an explanation lab only. That is, it provides an explanation of the steps you would follow to install an SMC in an actual setting.

1. Determine where your SMC will be mounted.

2. Mount the SMC to the wall studs by inserting the provided wood screws through the SMC's side panel. Center the screws in the slots for any future depth adjustments.

3. Tighten screws to secure the SMC to the wall.

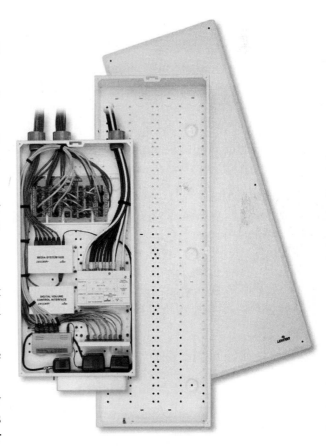

Figure 12-1-13 Leviton Series 280 Structured Media Center with modules affixed to the panel

4. Determine which SMC knockouts will be used for the ingress/egress of cables.

   Note: If cables are routing through the back of the SMC, break away appropriate knock-outs prior to mounting the enclosure.

5. Secure the SMC to the wall or to a plywood type backboard using wood screws through the slots on the back of the enclosure.

   Note: Securing the SMC to wall studs or using appropriate screw anchors is recommended to support the SMC's weight and internally mounted technology.

6. Establish an appropriate ground with one or more grounding screws that are located within the SMC. A minimum of No. 10, solid conductor, bare copper wire is recommended to ground the unit.

   Note: Proper grounding should always be verified by a qualified electrician and be compliant with the National Electrical Code (NEC).

7. Install the SMC cover. The hinged cover offers an additional lock.

Part 2: Identifying the SMC and Positioning the Modules

For this Part, you need to go to the lab wall, identify the modules, remove one of the modules, and reinstall it to the SMC using the push pins.

1. Select any module from the SMC and carefully pull on the push pins to remove the unit.

2. After the unit is removed from the SMC, determine where you want to place the module.

3. Align the module with the push lock pins on the distribution panel with the appropriate mounting holes on the enclosure. Seat the panel in place, resting flat against the back of the panel, and then press the push lock pins inward with your thumb to lock into place. You will know the push locks are secure if they "click" when pressed in.

4. After you have repositioned and securely seated the module, remove it and affix it to the location in which it was originally located.

Conclusions

During this lab, you learned how to install and mount the SMC to the wall studs. You identified the importance of placing the modules in their correct positions (horizontally or vertically) and demonstrated the correct way to affix devices to the SMC using push pins and/or screws. It is essential that you take cable routing into consideration, and allow sufficient space above and below each module for cable routing and power modules. Module placement is important so that your cable routing is neat and organized.

Exercises

What would be the disadvantage of mounting the advance telephone and video module on the bottom of the SMC panel?

Why should you keep the outlet cover in place for the unused outlets on the bottom of the SMC?

After affixing the AC power module to the SMC, is it recommended that a qualified electrician complete the wiring and grounding process?

Can the video modulator be mounted vertically? _____

What is the advantage of mounting the Decora media system hub horizontally?

_____

Door Facts

INTRODUCTION

When working in the field, you need to understand the different parts of a door. If you do not understand the terminology, there can be confusion about the proper installation and setup of security and automation products used with doors.

WHAT YOU WILL DO

Identify the different parts of a door.

Part 1: Definitions

The *hand* of a door refers to the way a door is mounted.

A *right-hand door* means that the hinge is added to the right side, whereas a *left-hand door* means that the hinge is on the left side.

The hand of a door is always determined from the outside. For an entranceway, this means the street side.

The *outside* of an interior door is the side from which the hinges are not visible when the door is closed, and the door opens away from you.

Handed (locks, and so on) indicates that the article is for use only on doors of the designated hand. If the wrong-handed lock is installed, the key will need to be inserted upside down (cut side up) to work, which will result in a shortened life for the lock.

Identify the door to the room you are in; is it left-hand or right-hand? _____

Is the outside of the door on the inside or the outside? _____

Part 2: Backset

When working with a lockset, you need to understand what the backset is: the horizontal distance from the face of the lock to the centerline of the knob's hub keyhole or cylinder (measured from the high side of a beveled door), as shown in Figure 12-2-1.

The backset on an exterior door is normally 2 3/4 inches. The interior door backset is normally 2 1/8 inches.

This measurement is very important when installing automated locks. If the incorrect lock is chosen, the door will not lock correctly.

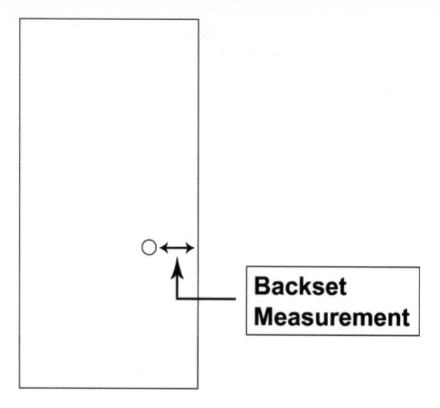

Figure 12-2-1 Determining the backset measurement of a door

Measure the backset of the door in your room: _____

Part 3: Door Locks

Door locks can be controlled remotely. In the event of failure, exit doors must fail so that the locks are open. Fire doors must fail so that they close automatically, but remain unlocked.

Remotely operated locking mechanisms need a power source. The power source may need to be a 16-gauge minimum, not the Ethernet wires used for other devices. This will be determined by the local inspector. Location of the power source depends on which hinge is set up to pass the power into the door.

Security doors have contactors that alert the security system when a door is opened. Contactors are usually two-piece devices. A wired device is mounted at the top of the door on the doorjamb. A magnetic sensor is mounted on the door, just below the wired device.

Why is the magnetic device mounted on the door? _____

Installing Jacks

INTRODUCTION

Jacks are used to access the permanent cable run in the walls. All wires should be terminated on the back of the jack. When dual jacks are used in a single faceplate, pairs of a four-pair cable should not be split between the two jacks. If dual jacks are used, two four-pair cables should be run.

Jacks come in a variety of categories: CAT 3, CAT 5, CAT 5e, and CAT 6. There are different categories for different performance requirements.

WHAT YOU WILL DO

Terminate and test a RJ-45 CAT 5e jack.

WHAT YOU WILL NEED

▸　Section of CAT 5e cable

▸　Cable stripper

▸　Cable cutter or electrician's scissors

▸　110 punch tool

▸　RJ-45 jacks

▸　Two CAT 5e patch cables

▸　LAN Pro Navigator

SETTING UP

Students will work in teams of two. One student will install a jack on one end of the cable; the other student will install a jack on the other end. Save this cable for possible use in future labs.

Part 1: Stripping

1. Use the CAT 5e stripper to remove two to three inches of the outer jacket. Figure 13-1-1 shows a cable stripper.

2. Separate the pairs from one another and fan them out, being careful not to untwist the pairs.

Part 2: Color Codes

Examine the jack closely. There are color designations on each side of the jack, which show where the individual wires will be terminated. The small *A* or *B* designates which wiring scheme will be used.

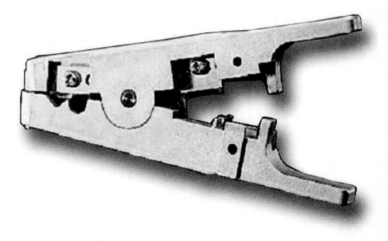

Figure 13-1-1 Cable stripper

The EIA/TIA 570A standard for residential cabling designates the T568A wiring scheme, which is what will be used for this installation.

The standard calls for the minimum amount of jacket to be removed, and no more than one-half inch of untwist for each pair.

Part 3: Termination

1. Place the end of the cable jacket just into the rear of the jack.

2. Place the white/brown pair into its appropriate position.

3. Use the 110 punch tool to seat the wires into their slots and cut off any excess wire.

4. Terminate the white/green pair in the same manner.

   Terminating the green and brown pairs first helps hold the cable in place and provides for the minimum amount of jacket exposure and the minimum amount of untwist.

5. Terminate the white/orange pair next, followed by the white/blue pair.

6. Have your instructor inspect your work.

7. Write down the maximum amount of untwist in your pairs: _____

8. Snap the dust cap over the completed termination (see Figure 13-1-2).

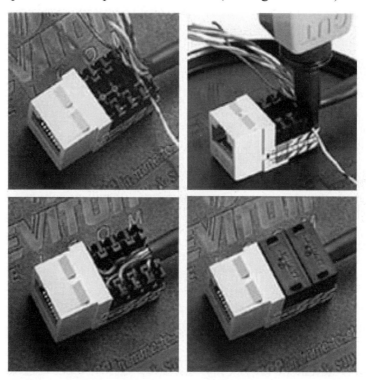

Figure 13-1-2 Termination process

Part 4: Testing

When both ends of the cable are terminated, test the cable using the LAN Pro Navigator. This will require the use of two patch cables. Figure 13-1-3 shows the LAN Pro Navigator.

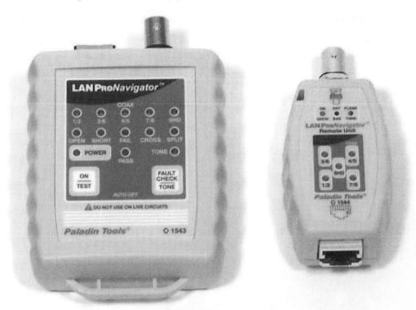

Figure 13-1-3 LAN Pro Navigator

1. Plug the patch cables into the LAN Pro Navigator and the remote.

2. Plug the other end of the patch cables into the jacks on the cable you just completed.

3. Turn the tester on and press Test.

4. Did your cable pass the test? _____

Work Area Outlets

INTRODUCTION

The main difference between a Level 1 and a Level 2 installation is the number of cables and types of cables run to a workstation. Usually, Level 1 is a two-port outlet with one RJ-45 jack and one type F coaxial connector. Level 2 uses a four-port outlet with two RJ-45 jacks and two type F coax connectors. Just because you are using a four-port faceplate does not mean that it is a Level 2 install. In this exercise, we will do a Level 1 install on a four-port faceplate; and instead of two RJ-45s and two type F connectors, we will install one of each and two audio connectors.

WHAT YOU WILL DO

Install jacks into a four-port outlet plate.

WHAT YOU WILL NEED

▶ One 12-inch section of CAT 5e cable

▶ One 12-inch section of series 6 coaxial cable

▶ One 12-inch section of speaker wire

▶ One RJ-45 jack

▶ One coax connector

▶ Two speaker connectors, red and black

▶ One four-port jack plate

Activity

1. Connect the RJ-45 jack to the section of CAT 5e cable by using the 568A wiring scheme.

2. Install an F connector on the small section of coax cable.

3. Look at the rear of the audio connector. Note that it uses a small set screw to terminate the audio cable. Loosen this screw, but do not remove it.

4. Strip about one-half inch of insulation from each conductor in the audio cable.

5. Twist the strands.

6. Insert one conductor into the red connector and one conductor into the black connector.

7. Tighten the set screws.

8. Before mounting the four-port connector plate, snap all the connectors into the plate.

9. Carefully feed the wires into the wall and mount the plate. Figure 13-2-1 shows a wall plate with the jacks installed on it.

Conclusions

Although this installation has four connectors in the jack, it is a Level 1 installation. Level 1 is defined by having a single CAT 5 cables and a single coax cable. Audio additions do not make this a Level 2 installation.

Figure 13-2-1 Wall plate

Testing Category 5e Cables

INTRODUCTION

Cabling is one of the most critical areas of residential networking, and the cable implementation is expected to last from 10 to 15 years. The quality of cable and connections is a major factor in reducing network problems and time spent troubleshooting.

To ensure that the connectors have been installed properly, a number of very simple and inexpensive basic cable testers are available. They usually consist of one or two small boxes with RJ-45 jacks in them. If both ends of the cable are plugged into the proper jacks, the cable tester will test all eight wires and indicate whether the cable is good or bad. If any of the eight wires has a break or short, or is miswired to an adjacent pin, the cable is bad. The simple cable testers may have a single light to indicate this, or it may have eight lights to tell you which wire is bad. These cable testers have internal batteries and do continuity checks on the wires.

Cable testers test functionality, but do not certify performance. Certification meters are expensive and sophisticated meters that perform a variety of tests on the cable. The EIA/TIA standards require certification to be fully compliant with the standards.

The basic cable tester used in the labs provides a large amount of information about the accuracy of cable terminations and the condition of cables.

The most common fault found with UTP cables is simple miswiring. Construction accidents that damage the cable account for a number of problems as well. Table 14-1-1 summarizes the T568A cabling standard.

Pin #	Pair #	Function (10Base-T)	Wire Color	Used with Telephony	Used with 10/100 Base-T Ethernet?	Used with 100Base-T4 and 1000Base-T Ethernet?
1	3	Transmit	White / Green	Line 3	Yes	Yes
2	3	Transmit	Green / White	Line 3	Yes	Yes
3	2	Receive	White / Orange	Line 2	Yes	Yes
4	1	Not used	Blue / White	Line 1	No	Yes
5	1	Not used	White / Blue	Line 1	No	Yes
6	2	Receive	Orange / White	Line 2	Yes	Yes
7	4	Not used	White / Brown	Line 4	No	Yes
8	4	Not used	Brown / White	Line 4	No	Yes

Table 14-1-1 T568A Cabling standard

WHAT YOU WILL DO

> ▸ Use a cable tester to verify your cable or cable runs.

> ▸ Demonstrate understanding of the T568A wire scheme used by the Leviton line of home networking products.

▸ Inspect your wiring using appropriate tools.

▸ Learn to identify specific cable problems.

WHAT YOU WILL NEED

▸ Cable with RJ-45 connectors on each end for each student

▸ Paladin LAN Pro Navigator

▸ RJ-45 plugs and CAT 5e stripping and crimping tool

Setting Up

Prepare a one-meter (three-foot) section of UTP cable for each student.

Part 1: Operating the Tester

The Paladin LAN Pro Navigator basic cable tester provides a large amount of information about the accuracy of cable terminations and the condition of cables (see Figure 14-1-1). It consists of two units: the *control unit* and the *remote unit*.

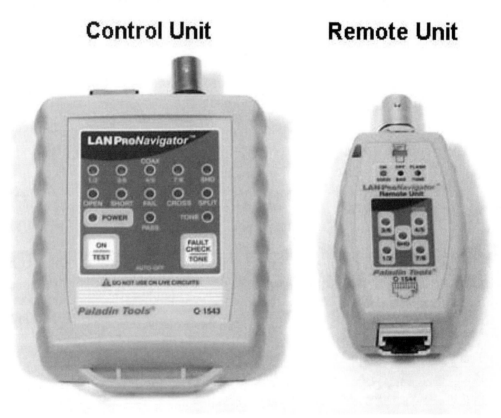

Figure 14-1-1 LAN Pro Navigator

The control unit and remote unit are equipped to test twisted-pair cables (shielded and unshielded), which connect with RJ-45 connectors. Both the control and the remote can test coaxial cables as well.

1. To test a cable, insert one end into the remote unit and one into the control unit.

2. Press the on button on the lower-left corner of the control.

   What happens to the power light?

3. Leave the unit alone for about 30 seconds.

   What happens to the power light?

   _____

   This feature ensures that the batteries will not run down if the tester is accidentally left on.

4. Press the on button twice.

   This will place it into test mode. Assuming that you are testing a UTP cable and that the cable is without fault, the first four lights on the top row of the control unit should begin to glow steadily, and the shield light at the right of the top row should begin to flash. This indicates that you have good connectivity in pins 1&2, 3&6, 4&5, and 7&8; that is, the cable is wired normally. There is also a Pass LED in the center of the bottom row that should be glowing. The remote unit, in the meantime, shows a red LED for each of the same pin combinations. Red in this case means good.

   When troubleshooting, the remote helps determine exactly what type of trouble is found.

5. Disconnect the UTP cable from the control unit.

6. Push the on button twice to place the unit in testing mode.

   The display on the control unit should be the same as in the previous example, in that all four pairs are being tested. A new LED should be lit in the middle row.

   What is the name of this LED?

   _____

   The cable fails the test (because it is not there), so the unit reads a failed condition as a result. The remote LEDs should all be dark. Dark in this case indicates no connection, or a bad wire.

7. Reconnect the cable you are testing.

8. Press the on button twice to put the unit into testing mode.

9. Press the fault check button on the lower right of the control unit.

   What happens to the display?

   _____

   The Fault Check button enables you to scroll through each pair individually. Pressing the button repeatedly will jump from pair to pair. After reaching the last pair, the tester tests the shield and then returns to the first pair to start the cycle over.

   The fault check button has one more important function.

10. Press the control button once or twice (pressing it a third time turns it off again).

11. Press and hold the fault check/tone button until the tone light goes on.

    This should take about 15 seconds. At some point, the pair that was selected should begin to flash. The corresponding pair on the remote should flash as well. This indicates that the unit's tone generator has been activated. You can cycle through pairs one through four, putting tone on each of them in turn. The fifth position causes tone to be switched down all the lines. Continuing to push the button starts the cycle over.

Part 2: Examining Cable Faults

The tester performs the following tests.

▶ **Open :** A broken connection in the cable or one of the connectors

▶ **Short:** A location at which one side of a pair touches the other side of the pair

▶ **Cross:** One conductor of a pair touches another conductor of the pair

▶ **Split:** One side of a pair is matched with the other side of another pair

▶ **Reversal:** The individual pair is terminated in the reverse order

▶ **Transposition:** A pair is terminated in the position of another pair

By using the fault check button, it is possible to examine each pair individually.

1. Cut one of the RJ-45 connectors off the end of the cable.

2. Install a new RJ-45 connector.

3. When installing this connector, reverse the blue/white pair.

4. Place the white of the white/green in place of the white of the white/orange.

5. Place the white of the white/orange in the place of the white of the white/green.

   Although it is unlikely that you would find all of these faults in a single connection, each of them is a common installation error.

6. Plug the cable into the tester and then test it.

   Notice that you must use the main unit and the remote to accurately determine the problems.

   Which LEDs are lit on the main unit?

   _____

   Which LEDs are lit on the remote?

   _____

Construction-Related Faults

INTRODUCTION

The combination of the LAN Pro Navigator main unit and remote unit helps accurately determine the type of construction-related fault. Suppose that you are testing a cable and the main unit shows that a pair is open, but the remote unit shows that same pair has a short. A likely cause of this could be a staple or nail driven through the cable. The conductors are severed by the nail and do not touch the nail on one side, but touch the nail on the other side.

Most construction-related faults result in open pairs or a combination of opened and shorted pairs. Reversals and transpositions would not be attributed to construction-related faults.

WHAT YOU WILL NEED

▶   Two cables with RJ-45 connectors on each end for each student

▶   Paladin LAN Pro Navigator

▶   Cable cutter or electrician's scissors

▶   Scrap piece of lumber

▶   Hammer and nails

Activity

1. Plug a small section of CAT 5e cable into the tester. Turn on the tester to verify that the cable is good.

2. Drive two or three nails into the cable. Do this over the scrap lumber so as not to damage a table. Drive them through the cable, but not so deep that they will be difficult to remove.

3. With the nails in place, retest the cables.

    What are your test results? _____

4. Remove one nail at a time and retest the cable after each nail is removed.

    Do the test results change? _____

5. Take another section of CAT 5e cable. Plug it into the tester to verify that it is a good cable.

6. Using a cable cutter of an electrician's scissors, cut approximately halfway through the cable. Retest the cable.

    What are your test results? _____

7. On another position on the same cable, cut halfway through again.

8. Test the cable again.

    What are your test results? _____

9. Cut all the way through the cable.

10. Test the cable.

What are your test results now? _____

Conclusions

Construction-related faults have symptoms different from installation errors. Proper test equipment can identify these symptoms and aid in discovering the location and the nature of faults.

CATV for Multiple Dwelling Units

INTRODUCTION

Multiple dwelling units (MDUs) provide a special challenge for structured cabling systems. Service providers typically terminate their services at a single point in the building or complex. This is referred to as the minimum point of entry (MPOE). Distribution cables are then used to deliver service to the individual apartments. These distribution cables must terminate in publicly accessible areas on the various floors. From these termination points, individual feeder cables are run to the different apartments. A small Structured Media Center (SMC) is typically located in each apartment. Individual outlets are fed from the SMC in the apartment.

WHAT YOU WILL DO

Learn the various connection points for a cable TV (CATV, which originally stood for Community Antenna Television) distribution within an apartment.

Activity

Figure 15-1-1 shows a drawing that represents a two-story, 16-apartment building.

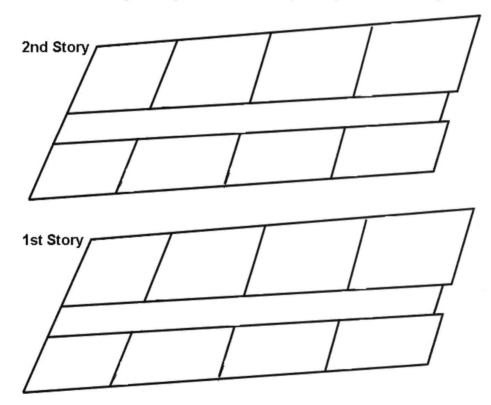

Figure 15-1-1 Drawing representing a two-story, 16-apartment building

1. Find a logical building entrance point and place an X at the MPOE location.

    This location will house distribution for that floor, and it will provide distribution to the floor above. The distribution cable to the floor above will usually be larger than series 6.

It may be one-half inch cable or RG11 series. Floor distribution from this room could be an amplified splitter if the runs to the rooms are long.

2. Draw a line from the MPOE to a likely distribution point for the second floor.

3. Place an X in the distribution point for the second floor.

   From each distribution point, a single series 6 coax will be home run to each apartment. Within each apartment a splitter is used to provide service to individual outlets.

4. Assume that each apartment has four outlets. Draw a schematic or sketch (for a single apartment) of the distribution system, beginning at an outlet and ending at the MPOE.

Wireless Communications

INTRODUCTION

In order for devices to communicate they must agree on a physical connection method such as wire, cables, microwave transmissions, radio waves, and even infrared. In addition, devices must agree on connection logistics such as data formats, speed, and syntax. Because the purpose of the majority of these devices is to make our lives more convenient, the trend is moving from wired to wireless technology.

This lab will help familiarize you with some of the popular wireless technologies available today.

Bluetooth

The Bluetooth standard uses radio frequencies to allow all types of devices such as computers, cell phones, stereos, headphones, and a multitude of other devices to communicate. Like any communication mechanism, Bluetooth has its pros and cons. The advantages of Bluetooth technology are that is inexpensive and very easy to use. Bluetooth-enabled devices do not require you to do any special configuration. In fact, Bluetooth-enabled devices find one another and begin communicating automatically.

802.11x or WiFi

The 802.11 standard comes in variations known as 802.11a, 802.11.b, and 802.11g, to name a few. These standards call for two components. The first is an Access Point (AP). The access point is connected directly to your wired network. It acts as the go-between connecting wireless devices and the wired network. Additionally, each device, such as a laptop or PDA, must be equipped with its own wireless access card to enable it to communicate with the AP. Each standard is different in its capabilities, as summarized in Table 15-2-1.

Wireless Technology	Operating Frequency	Operating Distance	Speed	Security
802.11b	2.4 GHz	Up to 150 meters or more	Up to 11 Mbps	WEP
802.11a	5 GHz	Up to 100 meters or more	Up to 54 Mbps	WEP
802.11g	2.4 GHz	Up to 150 meters or more	Up to 54 Mbps	WEP

Table 15-2-1 802.11 Wireless variations

The goal of 802.11b was to provide an inexpensive mechanism to connect wireless devices to a wired network. Of the three technologies, 802.11b has a good range and travels well through walls, so it is a good solution for a home network.

The goal of 802.11a was speed. The 802.11a standard actually was developed after 802.11b for performance reasons. But with the increased speed came decreased operating distance. Because

802.11a operates at a higher frequency, it does not travel as well through walls, so it is more limited in a home environment.

The goal of 802.11g is to provide performance comparable to the 54 Mbps of 802.11a while maintaining compatibility with 802.11b (and similar range as well). This compatibility is maintained because 802.11g operates in the same 2.4 GHz frequency as 802.11b. So, 802.11b and 802.11g devices will be able to communicate with each other, but when they do, the 802.11g will be no faster than the 802.11b product it is working with—they'll both be at the slowest speed common to each.

Wireless Security

The 802.11 standards implement security using two mechanisms called Wired Equivalent Privacy (WEP) and a special community name called an SSID (Service Set Identifier). The SSID can loosely be compared to a workgroup. Devices that share the same SSID can communicate. Additional security is added through WEP. WEP uses 40- or 128-bit encryption before transmitting data. This means that if your data is intercepted, it cannot be easily read.

In the majority of cases, for a typical home network 2.4 GHz will be the way to go, given its combination of good speed, range, reasonable cost, and upgrade potential. If you absolutely need higher speeds, a 5 GHz WLAN will do the job, but you'll need to factor in not only the significantly reduced range, but the fact that the signal may be excessively absorbed or reflected in the interior of your home.

Infrared (IRDA)

A third common type of wireless communication, which has been around for quite some time now, is infrared. Infrared, or IRDA as it is often called, is used commonly in television and electronic remote controls. Recently, it has been adapted for data transfers between notebook computers and personal digital assistants (PDAs). Infrared data transfers (between laptops and handhelds) have a maximum speed of up to 4 Mb/s and have an approximate maximum range of about two meters. The downside to IRDA is that it is a line-of–site technology. Devices must be aimed at one another to exchange date. IRDA transmissions do not pass through walls, so they cannot be used in different rooms of your home.

WHAT YOU WILL DO

▸ Identify implementation considerations for a variety of wireless technologies.

▸ List pros, cons, and implementations of popular wireless technologies.

OPTIONAL ITEMS

▸ www.bluetooth.com

▸ www.irda.org

Part 1: Understanding Bluetooth

Bluetooth devices communicate on a frequency of 2.45 gigahertz, which has been set aside by international agreement for the use of industrial, scientific, and medical devices (ISMs). Many common household devices such as garage door openers, baby monitors, and newer cordless phones all use the same frequency. The HTI should make sure during the design process that these other devices don't interfere with Bluetooth devices.

1. Name two advantages of using Bluetooth technology.

   _____

   _____

2. Name two disadvantages of using Bluetooth technology.

   _____

   _____

Part 2: Understanding 802.11 Technologies (b, a, and g)

1. What two pieces of hardware must be used to implement 802.11-based networking?

   _____

   _____

2. If you are concerned with the maximum range of your wireless signal, which 802.11 technology would best fit your needs?

   _____

3. If you are concerned with the maximum speed of your wireless connection, which 802.11 technology would best fit your needs?

   _____

4. If you want to upgrade the speed of your existing wireless network in the most cost-effective manner, which 802.11 technology would best fit your needs?

   _____

   _____

5. Explain the need for WEP technology.

   _____

   _____

6. When setting up a wireless network, each member of the network must share the same

   _____.

Part 3: Understanding Infrared Technologies

1. Why would a personal digital assistant (PDA) benefit from using IRDA technology?

   _____

   _____

2. List two limitations when using infrared technology to transfer data.

   _____

   _____